THE ILLUSION OF DEATH

by

John P. Dworetzky

HEARTHFIRE PRESS
Bellingham, Washington

Published in the U.S.A. by Hearthfire Press
Bellingham, Washington

ISBN: 978-0-6152-0549-6

Library of Congress Control Number: 2008927991

Cover design by Karen Casto

Illustrations by Joe Riehl

Front cover: Night train photograph © by *blackRose* All rights reserved. Available on Flickr at http://www.flickr.com/photos/rosi-sch/ and used with permission.

Back cover: Flames background vector illustration – iStockphoto (by avarkisp, 2006). Images obtained via iStockphoto were purchased legally and are used in accordance with their content license agreement.

Figure 2, 3d brain model - iStockphoto (by Max Delson, 2007)

Figure 8, from *The Mentality of Apes* by Wolfgang Kohler

Figure 9, The Milky Way's Center
Credit: NASA, COsmic Background Explorer (COBE) Project

Figure 10, Hubble's deepest view of the universe
Credit: Robert Williams and the Hubble Deep Field Team (STScI) and NASA.

Figure 26, Astronomy Picture of the Day, 2003 December 16 - Retrograde Mars (made by Tunc Tezel). Used with permission.

I hold to no religion or creed,
am neither Eastern nor Western,
Muslim or infidel,
Zoroastrian, Christian, Jew or Gentile.
I come from neither the land nor sea,
am not related to those above or below,
was not born nearby or far away,
do not live either in Paradise or on this Earth,
claim descent not from Adam and Eve or the Angels above.
I transcend body and soul.
My home is beyond place and name.
It is with the beloved, in a space beyond space.
I embrace all and am part of all.
—Jelaluddin Rumi (1207 - 1273 A.D.)

What if God was one of us
Just a slob like one of us
Just a stranger on the bus
Trying to make his way home
—Eric Bazilian, (1953 -   )

FOREWORD

If you are worried about dying, never fear. There are thousands of books that assure you that there is life after death, perhaps amid the delights of heaven, the torments of hell, or in other venues envisioned by various religions or philosophies. These books reassure millions of people, but John Dworetzky wrote this one for those who would *like* to believe in an afterlife, but who are not comfortable accepting on faith the immortality promised by religion and philosophy. If you prefer to base your conclusions about life and death on critical thinking, logical reasoning, and scientific research, or if you would like to base your beliefs on more than faith alone, you will find this book fascinating.

As John's friend and colleague, I know that he wrote it for you, but also to satisfy his own scientific curiosity. He wanted to know whether or not there is scientifically respectable evidence for the prospect and possibility of immortality, and long before he was diagnosed with the disease that took his life, he became convinced that there is. By the time you read the last page, you may well agree with him. The evidence that John so skillfully assembles from physics, psychology, mathematics, and other disciplines supports two main conclusions. The first is that death as we normally think of it truly is an illusion; it is highly unlikely that "when you die, you die, and that's that." The second is that though there is no logical way to predict what, exactly, happens to each life, to each consciousness, after death, the next step is guided by the laws of physics and chance, nothing else. In other words, John's mother may have been right when, on her deathbed, she told him "Don't worry; I am not afraid of death. I see it as a big adventure."

John was not immediately convinced. "I loved her dearly," he said, "but I thought that was about as nutty a thing as I had ever heard." In his final year, he worried that you would think the same of him, his evidence, and his arguments. With his trademark sense of humor intact to the end, he imagined approaching you as follows: "Hi, you might not know me, but I'm the first human being to actually figure out what happens to us when we die. The really good joke is that I am probably right and will never know it."

I wish you could have met John, but you can to get to know him a little now by following the logic and enjoying the humor that this labor of love contains.

—Doug Bernstein
Professor Emeritus
Department of Psychology
University of Illinois

PREFACE

John Dworetzky, the author of this book, died three weeks after completing its final editing, but before he could write the preface. As his wife, I have been present for many of the events that set the stage for him to write it, so I will have to sadly substitute for him in this endeavor.

I met John in 1973 when we were both students at Utah State University in Logan, Utah. When I first spoke with him he had an aura of sadness about him, though he also had much of the *joie de vivre* of a graduate student about to complete his degree. In the previous four years, he had lost first his father to heart disease, and then his mother, less than two years later, to cancer. They were each in their mid-forties at the time of their deaths, and this double tragedy would have a strong, long-lasting influence. John's paternal grandfather also died tragically of a brain tumor in his mid-fifties, accompanied less than a year later by his younger wife, from tuberculosis. We jokingly called this "The Dworetzky curse," which I proclaimed to be immune from because I kept my maiden name after I married John, as was beginning to be the fashion.

Each of us came from relatively religious backgrounds, but we shared a belief system that was more agnostic, or non-believing, than our backgrounds would normally foster. We also shared a strong belief that life itself was magical, that humor provided its sustenance, and that our minds could solve any problem, large or small.

As the death of his parents became more distant, John began to suffer other losses, first his high school sweetheart, then a beloved colleague, then his best friend from high school, and then other friends had their own close brushes with death due to accident or illness. By this time John had completed his Ph.D., taught college classes in psychology and child development, and written three college textbooks. But, the deaths of his parents and his friends, and then the birth of our two sons, continued to push him to not only ponder his own future death, but to figure out what might actually happen to his consciousness, his identity, after death.

Because of his training in psychology, he had a unique take on this thing called consciousness, and because of his interest and broad-ranging reading in modern physics, he had the knowledge to put the two disciplines' frontiers of thought together. The result is the book you now hold in your hands. I hope it provides you some comfort, just as it did to John in his own final days of life.

John had completed the first draft of this book by December 2005, when he found out, quite by accident, that he had kidney cancer. By January 2008, as John sensed the end was near, he spoke of a "night train" that would pass by every night while he was sleeping. Sometimes it woke him up to provide him with a new "cancer surprise," and sometimes it would just roar by. One night his father was driving and beckoned him aboard, but he resisted, as he still had some editing to do on this book, and he wanted to spend more time with me and with our sons. And so, first unable to go upstairs, then unable to walk, then confined to bed, he edited his book one last time and continued to crack jokes with the hospice nurses, sweet-tempered to his last breath on February 26, 2008.

—Karen Casto, March 7, 2008
Bellingham, Washington

CONTENTS

PROLOGUE

Each of us has a very strong commonsense understanding about the experiences we call life and death. We are born, we live, and eventually we die. Our lives follow an arc through time. Although we might have many experiences in common, we are all individuals.

However, if we take what is known at the forefront of science, we can come to a very different understanding of life and death than is dictated by our common sense, an understanding that leads to a surprising new world. Not only is this new world consistent with the recent discoveries of science, it is also a world in which many of the ancient paradoxes of the philosophers vanish. This new world is also a simpler world than the one in which we imagine we live, giving it, I believe, an even greater ring of truth.

However, this new world is so extraordinarily counterintuitive that I believe the best way to begin this book is to tell you the ending. You are immortal; just you and no one else. In fact, other people don't really exist. They are merely previews of other lives you will lead. In fact, nothing in your experience actually exists unless you are looking at it or thinking of it. You are the entire universe. Time does not flow forward, but rather flows randomly. During this random flow of time, you will experience every life that it is possible, both good and bad, because when you "die" you simply become someone else at some point in his or her life, a life that, with all its associated memories, now becomes yours and in a very real sense, was always yours. It is also an amoral world, because nothing you do, no behavior in which you engage, really matters. Acts and consequences are not actually connected in any way; the belief that they are is an illusion.

I am well aware that these assertions will seem fantastic in the extreme. But I can lead you to this world and do so with only reason, logic, and scientific discovery, and without the requirement of odd entities (no ghosts or spirits here) or strange concepts (no request for you to "head toward the light" will be made). I am a psychologist and I hope, a serious scientist, who has spent years studying human consciousness, and who also has an amateur's love for physics and astronomy. To be honest, the path down which I will take you has been taken by many others, mostly physicists and philosophers, but not with the conceptualization of consciousness I will bring to the discussion. That last bit, I believe, takes the final step into the world I will describe. So, let me show you what I have found. I can't prove it, as it lies beyond any possible proof. All I can show you is that your life, and what it actually *is*, might be amazingly different from the commonly held view that you are an individual who is born, lives, and dies. Come with me and determine for yourself what worlds may be.

THE AFTERLIFE

The yearning for an afterlife is the opposite of selfish:
it is love and praise for the world that we are privileged,
in this complex interval of light, to witness and experience.
—John Updike

In 1970, a shallow grave was discovered along the western bank of the Brazos River between Waco and Lake Whitney, Texas. In the grave were the remains of an adult and a juvenile. Both bodies had been deliberately covered over with large stone slabs. Found beneath the head of the adult, where they had been carefully placed, were seashell beads, knapping tools, red ochre, flint, and perforated canine teeth. Accompanying the juvenile was a small-eyed needle made from animal bone.

A more detailed analysis of the gravesite revealed that these two individuals had been laid to rest in or about the year 7965 B.C. This clearly wasn't a matter for the police.

What most captured my interest about this grave, however, was not its age or its occupants, or even the circumstances of their deaths, but rather the objects buried with these people. They were all things that could have been useful to the living and yet they were given up to the dead. But this is hardly unusual. It is quite common to find ancient burial sites filled with objects and things that might have been useful to the living. Ancient Egyptian tombs often included caches of gold, food, and even large boats; things that one would imagine could be of no use to the dead. Even today, it is not uncommon for people to place notes or favorite items with someone they love as that person is laid to rest. There is a feeling that somehow these things might be useful, or in some way comforting in a possible life to come.

Joseph Campbell, among other prominent students of mythology, believed that the concept of rebirth came to people from their experience with the natural cycle of the seasons. Winter, to be sure, was a time of death. But winter was always followed by spring, a time of rebirth. The planting of seeds in the spring, and what subsequently sprouted from them, also helped to establish the concept of rebirth.

To ancient peoples, seeds must have seemed to be magical things: Little hard bits of clearly inanimate stuff that, when planted down into the warm earth, would bring forth life anew. Things die, they go down into the earth, and then they return once again. People die, they are buried, and perhaps they, too, are reborn. Perhaps those who die will need certain useful things if they are to be reborn. It would seem reasonable, so why not send those things along with them when they are buried? Even if one doesn't believe that the objects or items will make the transition to the next world, at least one has made the effort, and the departed might in some fashion appreciate it.

The religions of the world commonly build upon some of these ancient ideas. They typically espouse an afterlife, a life after death. The afterlife might be a reincarnation, or a continuance in another place beyond this world; there are many variations.

But now, in the 21st century, many turn away from spiritual answers to the great question "What happens to us after we die?" and look toward science for a possible reply. Science has uncovered the processes of many phenomena once thought to be magical in nature, so it would seem to be a likely place to turn. For instance, science has shown us that a seed is nothing magical. It is a container for DNA, and it will replicate under the right conditions and bring forth a particular form of life. Seeds can even be manipulated to suit our needs. But concerning death, science appears to tell us little more than when you die, well, you die. Your brain turns to dust, more or less, and everything that was you is gone. When you're dead, you're dead, and that's that. Those who mark your passing can go ahead and bury money, food, and even boats with you if that's what they feel they need to do, but the fact is, "you can't take it with you, so why pack it?" But is this what the logical and rational process of modern science really tells us...that when we're dead, we're dead, and that's that? It might appear so. What other logical or rational alternatives could there possibly be?

But as I have indicated, reason, logic, and empirical observation actually indicate that there are alternatives, and that these are, in many ways, far more reasonable possibilities than, "when you're dead, you're dead." So, I shall begin with a discussion of an afterlife.

An afterlife isn't really what this book is about, however. If you are the only one alive and immortal, an afterlife will have little meaning for you. In fact, "after" implies the directional flow of time and I will argue that time doesn't work that way (usually). But the new world I wish to show you is so difficult to describe that I find the best way is to begin along a certain path. The only point to discussing any possible afterlife is to lay a foundation upon which further ideas can be built. This book is written so as to take you along a route that will pass through many doors, behind which will be almost always something astonishing, but which will have required approaching from a familiar direction before further understanding can be achieved. For this reason, I shall have us initially delve into topics which are by their nature preparatory, rather than central. An afterlife is one such topic.

Forgive me if I begin this discussion with some personal experiences and observations.

« »

When I was about three years of age, someone, and curiously I don't recall who it was, informed me that all living creatures eventually die. There were no exceptions, no corollaries exempting humans, none excusing me. My birth, it appears, had placed me in a situation that I couldn't get out of alive—exactly the sort of circumstance that I try hardest to avoid.

I have also had to tell my own children about the eventuality of death, which I suppose is a parent's responsibility. When telling my children, I tried to couch the news with, "but that probably won't happen until you're very, very old." That seemed to help some. Even so, they now regard me with a certain amount of suspicion that was absent prior to the telling.

I don't know about you, but when I was young I relied heavily on the "It won't happen until you are very, very old" part. But now, with the passing of time, what passes for very, very old no longer seems so old to me. Still, the fact remains, regardless of what sort of a happy face anyone might wish to place on it, we are all doomed. This, of course, begs the question: Is there anything that we can do about the mess we've gotten ourselves into?

One way to avoid everything coming to an abrupt and unwanted end is to postulate the existence of an afterlife: That is to say that we don't really die, even though to the uninformed it looks as though we do. Well, all right, maybe we die, but then we immediately go on to a new life, ergo, *after*life. And, just so that we don't feel too upset about the dying part, the new life will be a better life (unless you were a bad person, in which case you deserve whatever it is that is in store for you).

At a very basic level this pretty much is the promise of a great number of the world's religions. There is something waiting for us on the other side of life, something good. Or, conversely, we will be reincarnated as a gnat, or are heading to Hell—here I am reminded of the wag who requested that he be buried face down so that he could see where he was going. Either way, the point being made is that death is not an end; it is a beginning.

If you subscribe to any religion that includes an afterlife in its teachings, and you are a believer, well then, you are all set and are not doomed, assuming that the religion to which you subscribe is correct in its assumptions—I can't help but think of the 19th century British skeptic who suggested that all churches have carved in stone above their entrances, "Important, If True." On the other hand, if you are like me and are a nonbeliever, well, you are doomed.

I have come to prefer the term nonbeliever to atheist, as the latter seems a harsh word that to some might imply an anti-religious stance that I have not adopted. Although when pressed by a proselytizer to accept a pamphlet I might be overheard to make a caustic comment such as, "No, thank you, I don't believe in gods or devils," as a rule I am not a card-carrying atheist. It's just that, try as I might, I can't bring myself to believe.

I know that I am not alone in my non-belief, since demographic surveys show that roughly 10 - 37% of the U. S. population (and, I assume, a goodly portion of the rest of the world) is comprised of non-believers, depending on how one phrases the questions asked about belief. I also suspect that there are quite a few closet non-believers, inasmuch as they give lip service to their religious affiliation, but secretly harbor deep suspicions or reservations about its validity.

It is perhaps odd that I am a non-believer since I come from a family that certainly had its share of believers. This leads me to consider that my non-belief isn't genetic in origin. Perhaps the cause lies in my environment—it might be worth a moment's consideration.

My father's father was a Jew. He emigrated from Russia to the United States in 1905 to escape the various pogroms that were in vogue, and also to avoid being conscripted to fight the Japanese, a people with whom he had no personal disagreements. My grandfather's solution to this predicament, reasonably, was to place the Atlantic Ocean between himself and the Tsar.

My grandfather was industrious and graduated from medical school to become a tuberculosis specialist. One of his patients was a petite Irish Catholic woman with whom he fell in love and promptly married. Believing that God was trying to tell him something by sending him my grandmother, he converted to Catholicism, at which point most of his brothers and sisters pretty much decided that, as a now non-Jew relation, he was dead to them (my late Great Uncle Leon was a kind exception).

My father was born and raised as a good Catholic, but he was proud of his Jewish heritage. He, quite naturally, married a Protestant.

I was raised in Brooklyn as a good Catholic-Protestant-Jew. I chose to attend parochial school so I could be with my best friend, who was wholly Catholic (no pun intended). But just to be certain that the nuns weren't getting the upper hand, my mother made sure that I knew of Martin Luther, and of the Pope's failings. My many Jewish friends and their families also shared with me their views, beliefs, and celebrations.

My father was a physician. Although he was also a Roman Catholic, or at least that was what he told people, he had a tendency to ask unsettling questions concerning The Faith. I recall once mentioning the concept of "intelligent design" (despite the recent press, it is hardly a new idea), and telling him that I had learned that day in school from Sister Frances-Julie that, "The existence of a watch presupposes a watchmaker; therefore the existence of a man presupposes a man-maker." My father glanced over at me and said, "I see, well then, the existence of God must presuppose a god-maker." I remember going to bed that night wondering if Supergod had made God? If so, it would only seem reasonable to assume that Hypergod had made Supergod, and that Ultragod had made Hypergod. Someone must have made Ultragod, but I wasn't sure who, since I had run out of prefixes.

Although I had Christian parents, I was also grateful to have Jewish friends and a Jewish heritage. Occasionally, this arrangement led to a certain amount of confusion. At Christmas services the nuns thought it unfunny when I remarked that it sounded like they were singing, "Oy vey, Maria," or when I referred to the head nun as "Mother Shapiro," rather than Mother Superior, a mistake I still claim to have been an honest one.

Eventually I got things straightened out and, quite naturally, married a Mormon (a lovely young woman I met while attending graduate school in Utah). Other than occasionally arguing that I can rightfully avoid work during any religious holiday that might make an appearance on the calendar, I am, as I said, a non-believer. I have nothing against those who do believe, and self-righteously defend to the death their right to believe whatever they claim.

I mention all of these personal observations and experiences because this book has absolutely nothing whatsoever to do with religion; I want to stress that fact. I do not approach its writing with a particular religious point of view, other than to have none. This book concerns itself only with one great question: "What is this experience we call life and what happens to us after we die?"

But for now, however, it will be easier to focus on the latter half of that question, namely "What happens to us after we die?" Just for now, let us assume this is the great question we are pursuing. I know this seems like a religious question, but I believe that there is a wholly secular approach to answering it

THE SECULAR VIEW OF DEATH

For three days after death, hair and fingernails continue to grow
but phone calls taper off.
—*Johnny Carson*

Among non-believers I find that the great question, "What happens to us after we die?" is typically met with variations of the same basic answer, namely, "When you die, you're dead, and that's that." Originally, I accepted this conclusion as not only rational, but the only logical possibility. I concurred that when you died, you didn't go anywhere. Hard as I might try, I couldn't bring myself to imagine a spirit that floated off somewhere, and I still don't. All I could comprehend about an afterlife was that once you died, you just lay in a box and rotted (unless, of course, you'd been cremated, in which case you smoked for a while and then got urned or scattered someplace). Logic dictates that when you're dead, that's it. You've kicked the bucket, you've bought the farm, you're pushing up daisies; you're making your last call from the horizontal phone booth. However you say it, that's all she wrote.

If, however, as previously stated, you are religious, then God bless you—you know what will happen afterwards and you're all set. The Mormons (or LDS for Latter Day Saints) even know what they will wear in Heaven, and know with certainty that they will be reunited with their families once they get there (leading a few Mormons I know to wonder if being reunited with their families will be evidence of their arrival in Heaven, or perhaps lesser regions.)

But I, and others like me who are non-believers or who have serious doubts, are simply out of luck. Apparently we're not going anywhere. My grandfather (my mother's father, not the doctor from Russia) said it best when we once attended a funeral of one of his business acquaintances who was an avowed atheist. The man was lying in his coffin dressed in his finest suit. My grandfather whispered to me, "All dressed up and no place to go."

So, if we are not going to look to religion to provide an answer to the great question, "What happens to us after we die?" where can we look? I am trained as a scientist, one who relies on empirical research and rational processes, so I find that I am unwilling to turn to any pseudo-science for help. Nor do the ideas of reincarnation common to some Eastern religions impress me. I am not saying that I know for a fact that that I won't return as a flea because of the poor life I have led, or that any religious belief for that matter is wrong (I am not so arrogant as to imagine that I have all the answers), it's just that I would like strong data to address the great question, or barring that, some evidence for an hypothesis that seems reasonable based upon scientific knowledge. But alas, there is nothing empirical. Anyone who requires empirical research to yield a solution, then, is forced to conclude that the answer to the great question, "What happens to us after we die?" is, in fact, unknowable. That is what I too believe; it is unknowable.

So what is the point asking the question? If I am not about to implore you to head toward the light, and the answer to the great question is, in fact, unknowable, then what's left to discuss about that question? If we are not going to turn to a religious answer, or about to posit some sort of ethereal "life-force energy" or some such thing that will carry our "spirit" off somewhere, what could possibly be the end result of dying, other than to be dead? This brings us back to, "When you're dead, you're dead," a dead end if ever there was one.

Perhaps, then, the best way to approach the great question, "What happens to us after we die?" is to examine the ubiquitous secular answer, "When you're dead, you're dead, and that's that." And that is where I wish to turn next—there is something curiously unsettling about that conclusion.

DEAD SO LONG

It may be he shall take my hand
And lead me into his dark land
And close my eyes and quench my breath—
...I have a rendezvous with Death.

—*Alan Seeger*

What does it feel like to be dead? Scientifically speaking, this question makes no sense whatsoever. When you die, your brain ceases to function. At that point all sensations and perceptions stop. Once you have ceased to exist, you can't "feel" anything. Oh sure, you don't completely cease to exist, in the sense that you leave a body behind—for a while, anyway, unless you are cremated—and then you leave ashes. I suppose that's a form of "existence." But sooner or later the body or ashes you have left will break down into their constituent atoms and molecules.

Five billion years from now, our sun will have grown into a giant red star as it cools and begins to fuse its remaining helium and carbon. By then the Earth will most likely be burned to a cinder by the sun's expanded radius and many of the atoms that had once constituted your body will be scattered about in space, much as they were billions of years before your birth.* This leads us to an interesting observation—the atoms that came to constitute your body 5 billion years *before* you were born will be in

---

* I have couched my comments about the Earth eventually burning up with the phrase "most likely" because chaos theory does provide for the possibility that the Earth, through a series of unfortunate and one-sided orbital perturbations (not unlike a flipped coin coming up heads 500 times in a row) might just decide to leave its orbit altogether and go wandering off into deep space and thereby avoid the sun's later expansion, and incidentally, end on a cold note rather than a hot one...although I am told that this is exceedingly unlikely and that there is, as of yet, no pressing need to stock up on firewood.

roughly the same state of scatter 5 billion years *from now*; truly ashes to ashes, and dust to dust. Another way of saying this is that you didn't exist 5 billion years ago, and you won't exist 5 billion years from now. In that sense, "after you're dead" in terms of physical chemistry, is no different from "before you were conceived." In both cases, you are non-existent. Clearly, if death simply returns you to your former state of non-existence and random scatter, then, interestingly, you've been "dead" before. That is, there will be no discernible difference between the "you" of 5 billion A.D. and the "you" of 5 billion B.C. In the future you will be dead, but you were similarly "dead" in the past.

It's a silly, but natural feeling that if you have been dead before you ought to remember what it was like. Let's see, what was the year 498,331 B.C. like for me? Oddly, I find it to have been very much like 903 A.D. Neither year had a single moment about which I could complain. On the other hand, I don't recall anything personally good about those years, either. My best recollection is that they passed by fantastically quickly. But I do know that I didn't mind being dead at the time. It didn't bother me in the least. In fact, even though I was dead for billions and billions of years, it didn't upset me—not one bit. I imagine that the years 2299 A.D. and 929,989,677 A.D. will be "experienced" by me in much the same way as were those earlier years.

Of course, these aren't real memories I am citing. I didn't exist during those years and neither did you, so there was no functioning memory into which we could place any real experience. It is interesting to think, though, that we were dead before and that it didn't bother us. Of course, that's because at the time there was no you or me to be bothered. I imagine that it will be much the same for both of us after we are gone.

But now consider how much of our time was spent being "dead." Look at Figure 1 and you will see a time line extending some 10 billion years in each direction from the present. I suppose that you could go backward or forward in time a lot further than that, but the physics begins to get a bit obscure. There might be a beginning or an end to time, but then again, there might not be. (In fact I have always found it interesting to consider that imagining an end to time is every bit as hard as imagining time never ending.) Perhaps time has always flowed forward, or perhaps it loops back on itself. There are many good mathematical models that show such possibilities. But for our purposes, it is not necessary to look too far in either direction.

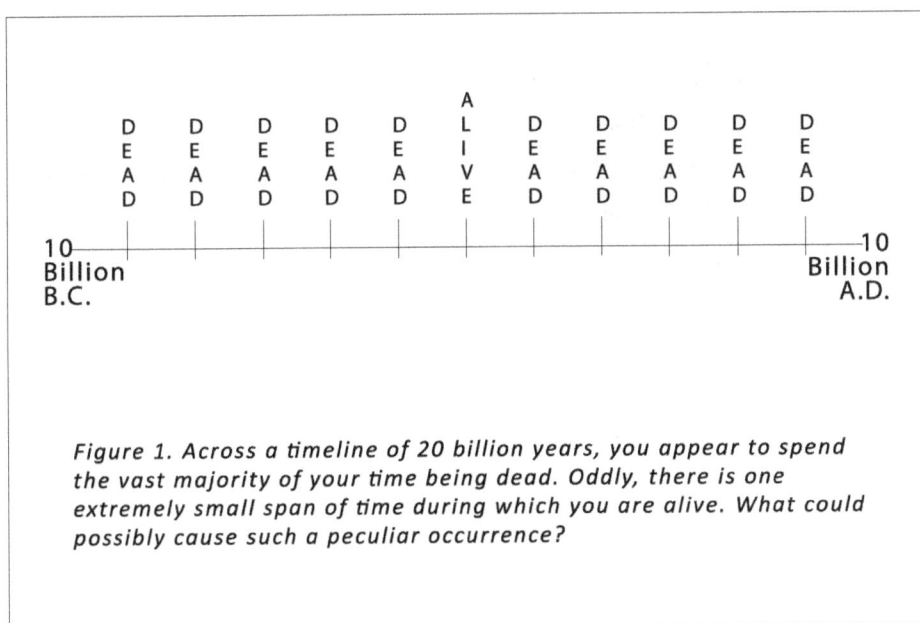

Figure 1. Across a timeline of 20 billion years, you appear to spend the vast majority of your time being dead. Oddly, there is one extremely small span of time during which you are alive. What could possibly cause such a peculiar occurrence?

What's interesting is how much of that time-line you spent being dead. If you start over on the left of the line and slowly run your finger along it moving right, you will pass over a huge number of years in which you were dead. You can just move right along, checking your status as you go—dead, dead, dead, dead, dead, dead, dead, dead, dead, dead, alive, dead, dead, dead, dead, dead—wait! What was that one place? In one little spot, one ridiculously short span of time, you were ALIVE. Then you very quickly reverted to your normal state of being dead. What in the world was that all about? That was certainly a strange thing to have happened.

In fact, your coming to life is such an odd occurrence that if I couldn't point to the existence of you as proof that you could come alive; it would seem too fantastic to ever be considered as a serious possibility. Somehow, in some way, this universe created you. Out of a universe that is mostly space and hydrogen came this planet and you. Good grief, what were the odds? What an amazing thing to happen! It doesn't last long, though. You are dead soon enough and your atoms go back into their more commonly arranged state (haphazard scatterment, for lack of a better term). But amazingly, there is that little ALIVE blip on the great time line in amongst all those dead spots. What to make of that? Wild, crazy, random chance, I guess. What a thing to happen. Yet, as a scientist it makes me very suspicious. I feel like a detective with 30 years on the force who sizes up a situation and, even though he doesn't know why, he can just tell that something's not right. I just feel it. Something strange is going on here, but what?

ACTUALLY, WHAT ARE THE ODDS?

Sometimes I've believed as many as six impossible things before breakfast.
—Lewis Carroll, "Alice in Wonderland"

The fact that you and I came alive in this universe strikes me as so amazing; I can't help but wonder just what really were the odds? Although this won't get us anywhere (as you will soon discover), it is still fun to attempt the calculation. Plus, doing so will reveal a concept that I do wish to introduce here at the beginning of our journey. So let's try to calculate the odds of being born and begin with a few basic common assumptions about how you came to be (assumptions that might or might not be true).

First the right sperm had to meet the right egg. Just having the right two people get together won't do it, which is why any siblings you might have aren't you. Well, millions of sperm race toward just the one egg—who hasn't seen that one in a biology or health class? So you won that lottery. And lottery is not a bad choice of words, either. The odds were tens of millions to one. So now you know why your life has been going the way it has been going. You used up all your good luck before you were born.

But wait (as they say in infomercials), there's more! It wasn't just that the right sperm had to meet the right egg (the one your mother ovulated that particular month), but your mother had to meet your father. How did they meet? How many millions of little things might have occurred to prevent that meeting or to keep them from wanting each other as much as they apparently did? Or wait, I just realized, before your father and mother could meet *they* had to be born. So here we go again with the millions of sperm and the egg, twice this time, once for each parent. And your grandparents had to meet and fall in love. How unlikely was that? Anyway, you can see where this is going. Your great-grandparents (all the millions of sperm and egg times four, plus they all had to meet and fall in love.) Well, I guess this goes on back to who knows when? The point is that to figure the odds of your being born will now require Stephen King's calculator. It will be well over a trillion to one, or perhaps a quadrillion, or a quintillion (these interesting number names actually go on—pentillion, sexillion, septillion, octillion, nontillion, decillion, undecillion, duodecillion...). Whatever the odds, they appear to be beyond astronomical! In fact, I would argue that they are so high that you couldn't possibly exist. And yet, you do. If you ask me, that is highly suspect.

Also consider the fact that the sperm cell that reached the egg that created you isn't really the sole determiner of your creation. For instance, identical twins are produced from the same sperm-egg union, which produces a zygote that splits after being formed, and yet they are different people. This, in fact, appears to make the odds of your birth even more fantastic, as it now appears that there are other variables involved in the making of you besides just the right sperm-egg combination.

It can't be, then, that you just got lucky and pulled off a vigintillion-to-one shot (there is such a number); the odds must not be as great as I am imagining them to be. How can we bring the odds down to make your birth a more likely possibility? I am not sure how, but it seems that we had better try.

To reduce the odds some, let's begin with the following observation: Have you ever sat in a room in which there are mirrors in front and behind you so that you could view the reflections of your reflection? Perhaps you've had this experience while getting your hair cut. You see ever smaller and smaller images of yourself disappearing into infinity, going on forever. While this concept might seem appealing, the trouble with it is that the reflections don't actually go on forever. Light waves, which carry the information of your reflected image, have a certain length. Visible light ranges in length from about 400 nanometers (400 billionths of a meter) to about 700 nanometers. The longer wavelengths are red, and the shorter, blue. As each successive image gets smaller, less light is reflected from it, and therefore, less information. Because of this, each successive image gets fuzzier and fuzzier as data are lost owing to the smaller size of the reflection. Finally, the image gets so small that it becomes smaller than the longer wavelengths of light and the red drops out, leaving the last image very, very fuzzy, and pretty darn blue. There is an image after that that is even fuzzier and ultra-violet (assuming that the mirror coating hasn't absorbed it) but you couldn't see that one, anyway. So there you have it. After a finite number of reflections the "endless" reflections come to an abrupt end. (Not to mention that for light to go on for an infinite number of reflections, even at the speed of light, it would take an infinite amount of time).

The same "dropping out" occurs when we consider your relatives (I mean nothing personal). Genealogy is fun and I have dabbled in it hoping to find either presidents or horse thieves, but coming up with no one so interesting. Still, it is fascinating to consider one's family tree. Like the "endless" mirror reflections that aren't at all endless, if you go back far enough in your family tree, strange as it sounds you will find that you are *unrelated* to many of your forbearers.

Consider that the average human being has a total of about 62,000 genes in each of his or her body cells (although current estimates range from a low of 27,462 to a high of 153,478, depending on how the decoded human genome is interpreted. But for sake of discussion, let's use the 62,000 figure. It will make no difference for our purpose.)

Roughly half of these genes came from your father, and the other half from your mother. Similarly, you received one quarter of your genes from each grandparent. If you go back 15 or 16 generations you will find at that distance that you received about one of your genes from each relative. If we define a "relative" as someone with whom you share chromosomal genes, than go about 17 generations back and you will find that half the members of that generation in your family tree are unrelated to you (inasmuch as they have not passed a single one of their genes down to you via a direct line of inheritance). Go back about 18 generations, and three-quarters are unrelated, and so on. In fact, many of us share similar or identical genes by random chance, and so if you only go back about 13 generations or so, you will find that your forbearers at that distance are no more "related" to you than would be a similar sized group chosen randomly from the phone book.

Actually, things are a bit more complex than that. About 60,000 of the 62,000 genes that go into making a human being are shared by all of us. These are the genes that order a head, or a pair of eyes, or that tell cells how to divide. In fact, about 30,000 of our "human" genes can also be found in a banana (or I suppose you could say that 30,000 banana genes are also found in humans, but I think it sounds better the other way around). Well, banana cells need to divide, too, and their cells share the same mechanism found in ours. (Perhaps this will also explain why you have *appeal*—please forgive me, but I have a weakness for awful puns).

So there are roughly only 2,000 genes that determine the most obvious differences *between* members of our own species. In fact, if we examine two people taken at random, the odds are that their genetic codes will only show a difference every 1,200 letters of code or so. The rest of their codes will be identical.

The various combinations of these genes are apparently responsible for the more than 22 billion people who have been estimated to have walked the Earth since the beginning of our species. That certainly seems possible. Imagine a deck of cards with 2,000 cards and you can guess how many different sequences of cards could be dealt, certainly more than 22 billion. Perhaps a clearer way of expressing that these 2,000 or so genes interact in a multiplicative fashion, is to point out that the more than 6 billion individual faces that exist in the current human population, and which enable us to quickly tell one person from another, could be the result of combinations of as few as just 33 different genes.

Although many of these possible genetic combinations might not produce a viable human being, (many of the genes we inherit appear to code for nothing[*]) there certainly seem to be enough combinations to account for all the people who have ever lived and perhaps a very great many more to come.

So, if genealogy is your thing, I would recommend that you don't go back more than about 13 generations. Once you have gone back that far, you will have found all of your related forbearers. After that, they are mostly strangers even though they might be sitting on a branch of your family tree.

So now that we don't need hundreds of generations meeting and falling in love in order to create you, we have the odds of you being born back down to, who knows, perhaps a few undecillion to one? Or perhaps there are less possible combinations that can yield a human being and we are really talking about the ever-so-more-likely trillion-to-one shot? Either way, I am still having trouble believing that either of us is actually alive, and my suspicions have not been allayed by these recalculations in the slightest. I still say something very peculiar is going on. I just have trouble believing in long shots that are that long.

---

[*] The process of evolution and natural selection would suggest that code might well evolve that had no purpose. It is assumed that such code was never selected against simply because it never got in the way, inasmuch as it never interfered with something vital.

Now, I know what you're going to say, that a trillion-to-one shot, however unlikely, is *still possible*. And, I agree with you. In fact, it is this inescapable logic that causes me to buy a lottery ticket each week even though I actually *can* do the math. They always say the one thing that hooks me, "If you don't play, you can't win." In accordance with this view, there are also a number of scientists and philosophers who like to point to what they refer to as the *anthropic principle*, namely that the only beings who could ponder how a trillion-to-one shot could ever occur, are the ones who had it happen to them. Well, perhaps so. But still, I am very suspicious of odds that range in the magnitudes that we have been discussing.

Okay, for the sake of argument, let's say that you and I and every other person on this planet won the super-duper celestial lotto. Whether we believe it or not, let's make that assumption and use it as a jumping off point. So, thinking back to that deck of 2,000 cards (and additional cards for whatever variables allow twins of the same genetic make-up to be different people), we can imagine that the deck got shuffled, the cards got dealt, and you and I came up along with everyone else who is now, or was once, alive. The rest of the combinations are out of luck, or perhaps they'll show up later.

Come to think of it, we can make our births more likely and lower the terrible odds that we have been discussing even further by imagining that perhaps there are only enough possible combinations to make a trillion different people, and since 22 billion have already been born that would make the odds of our births about 45 to 1. Or, even if a quadrillion different people were possible, that would only worsen the odds to 45,000 to 1, well within lotto-winning range. And, just because there are millions of sperm cells, it doesn't necessarily mean that only one would have led to you. Perhaps any of a few thousand in the bunch would have made you. Who is to say? We might even get around the problem of your parents having to meet and each set of your grandparents having to meet and so on, by hypothesizing that the creation of your conscious mind might have also been possible as a result of the union of different people than those to whom you were actually born. Just because your body might look different doesn't necessarily mean that it wouldn't be you, would it? Who can know? So perhaps that is how we both came to be. There are only a set number of possible people and once lots of them start being born, our individual chances began to increase until we got lucky!

But before we start to think that the odds are now being lowered into a more manageable range, we can once again raise them. What were the odds of a species appearing that would have the capacity for conscious awareness that humans possess? What were the odds of a planet forming in our solar system that could support such life? Planetary scientists tell us that without a huge world like Jupiter nearby to suck up much of the space debris in our solar system, so many large asteroids would have bombarded the Earth, and so often, that human life would probably not have arisen. Or, if the Moon had not been there to stabilize the wobble of our axis, seasons probably would have been far too extreme to support the development of human life. Or, if water did not have the very odd property of expanding when it went from a liquid to a solid (which is chemically very rare) ice would not float and the oceans would have frozen solid, making the development of advanced life quite unlikely. Or, if the Earth itself had been just a few million miles further from or closer to, the sun, the oceans would have frozen solid or boiled away. So what were the odds of a planet forming with all the characteristics of Baby Bear's porridge ("just right")? And what are the chances that our species line might have survived the many cataclysms that have befallen other hapless creatures (in fact, well over 99% of all species known to have ever existed are now extinct)? As Bill Bryson put it in *A Short History of Nearly Everything*:

We are so used to the notion of our own inevitability as life's dominant species that is it hard to grasp that we are here only because of timely extraterrestrial bangs and other random flukes. The one thing we have in common with all other living things is that for nearly four billion years our ancestors have managed to slip through a series of closing doors every time we needed them to.

So you see the odds of us ever being born are once again worsening. As you can probably tell by now, the odds game is an impossible one, since we will never know all the variables that might be involved. Try as we might, we will never know if we are talking about odds 10 to 1 in our favor, or a vigintillion to 1 against. I am guessing that they must not be too high, since we are here, but who can say? Perhaps we did get fantastically lucky. While odds are interesting to think about, considering them appears to have gotten us nowhere in our quest for a better understanding of the great question, "What happens to us after we die?"

COMBINATIONS OF "YOUNESS"

The task of science is to stake out the limits of the knowable,
and to center consciousness within them.

—Rudolf Virchow

While we were discussing the odds of being born, you might have noticed that we were contemplating an interesting concept. The concept is that a particular combination of genes, or chemistry, or *something*, leads to the creation of particular people. Actually, this is a very common assumption among lay people and scientists alike, and it makes some sense. Although people are very similar in many ways, no two people are exactly alike, not even identical twins. Identical twins have different fingerprints, different iris patterns, and other clear differences. So it seems quite natural to ask how it is that you came to reside inside your body, looking out through your eyes, listening through your ears, while everyone else is—well—everyone else. How come you are in your body, while other people are in theirs? At the risk of sounding ungrammatical, what exactly is "you?" What causes your "youness?"

The most common answer offered by the few who spend time dealing with such questions, at least the most common modern answer, is that it is based upon individual differences. You have your own fingerprints, your own iris speckling, and, well, your own consciousness. Somehow, something has caused you to have your consciousness and not someone else's. Perhaps there is a certain biochemical/genetic code that yields a YOU, while other codes yield a HIM or a HER. It certainly seems reasonable. Other people are different from you, and you are not they; so there must be some fundamental difference between you and them.

So now the question becomes, just who exactly are "you?" And, if who you are is the result of some sort of unique biochemical combination that has obviously occurred once (since you are now alive and reading this book), what's to stop it from reoccurring? Furthermore, if that combination did reappear, would that be "you" once more, alive again?

Such a suggestion is not new to philosophy, psychology, or even the mass media. People such as "Bridey Murphy" and others have laid claim to recollections of past lives. There was even a movie made using that concept as a plot line called *On a Clear Day, You Can See Forever*. But is such a thing possible? Does it make any logical, rational, or scientific sense? Could you be reborn? Did you live past lives? In place of "afterlife" could we simply substitute "next life?"

If we make an attempt to look at this interesting problem rationally, we will run into three great issues. First, what exactly is the "you" that is being recreated, and is it really you? Second, and I shudder to mention it: What are the odds? And for the third, and this is the really wicked one, we will come face to face with the paradox of supervenience.

ON A CLEAR DAY, HOW FAR CAN YOU REALLY SEE?

And as to you Life I reckon you are the leavings of many deaths,
(No doubt I have died myself ten thousand times before.)
—Walt Whitman, "Leaves of Grass"

In the United States in the early 1960s, there was a book, and then a TV series, called *In Search of Bridey Murphy*. It was the true story of a modern day woman who claimed to have recollections from a past life. She recalled having lived in Ireland in fairly vivid detail, and said that her name had been Bridey Murphy and that she had died during the great potato famine of the 1840s. A number of her recollections appeared to match real occurrences.

Many others have had similar past life experiences, including actress Shirley MacLaine and General George Patton. Some have posited that there is a life-force energy that revisits different people throughout the ages and that this might account for their shared memories.

But please, do not be alarmed. I am not about to argue that such anecdotal evidence is useful proof of past lives. In fact, I don't believe that any of these people, well intentioned as they may have been, was actually "remembering" anything, other than perhaps their own imaginations disguised as recollections. I am also not above believing that some of those claiming to recall past lives are out and out frauds. One of my personal heroes, The Great Randi (magician James Randi), has done yeoman's work spending much of his talent and career debunking and showing up many of these frauds and would-be tricksters.

Furthermore, I promised to rely solely on logic, reason, and scientific knowledge in pursuit of the current question under discussion, "What happens to us after we die," as well as all other questions, and I will not take the cheap way out and turn to mysterious life-energy forces unknown to man (although who's to say what is out there remaining to be discovered).

If we use reason, logic, and scientific information, then it is possible to deduce that there is a very high probability that none of these people with so-called past life experiences is remembering anything from a past life. This becomes abundantly clear when we consider what human memory actually is, so let's take a few moments to do exactly that. Also, by examining memory, we can better appreciate how the human brain works and perhaps get a glimpse of where "you" might reside within its many folds and convolutions. That, in turn, might actually lead us somewhere.

Memory is a biochemical way of storing information in the brain. Although there are many differences between humans and computers (e.g., humans have neither a CPU nor "Intel inside"), it is not wholly a bad analogy to compare human memory with computer memory. Both require an input mechanism, storage capacity, and a way in which to retrieve the data stored.

Particular parts of the human brain appear to be especially involved in these differing aspects of memory. Damage to any one part of this complex interacting biochemical system can lead to curious and peculiar memory deficits. By examining a few case studies, we can highlight how human memory is dependent upon various parts of the brain.

First, let's consider the case of Mrs. Duke, a 66 year-old woman who was brought into the hospital by her husband, who said that his wife's memory was lapsing and that she needed help. On call that day in the hospital emergency room was Dr. Tony Dajer. After introducing himself, Dr. Dajer began to examine Mrs. Duke for any obvious physical problems. After completing an initial examination that failed to show anything, Dr. Dajer decided to conduct a simple memory test.

He told Mrs. Duke, "I'm going to name three objects. In five minutes I'll ask you to repeat them back to me, all right?"

Mrs. Duke smiled and said, "I'll do my best."

Then Dr. Dajer looked around the room and saw three objects he could name and said, "Bed, chair, sink. Got it?"

Mrs. Duke answered, "Bed, chair, sink, yes."

Dr. Dajer knew that anyone whose memory was functioning at a reasonable level would be able to recall all three items easily after only a 5-minute delay. If Mrs. Duke could not recall one of the items, he reasoned, it might mean that she did indeed have a memory problem.

After five minutes, Dr. Dajer returned. Mrs. Duke seemed in good spirits and well enough, so Dr. Dajer said, "Well Mrs. Duke, what were the three objects?" That was when Mrs. Duke looked at Dr. Dajer and asked, "Who are you?"

To say that Dr. Dajer was shocked would be an understatement. After further examination, Dr. Dajer realized that all Mrs. Duke had to go on when dealing with *new* experiences was what is sometimes called "short-term memory." Mrs. Duke could recall things from her more distant past well enough, but she could not remember anything new she encountered for more than about 30 seconds after it was out of her sight! In fact, she had to be constantly told where she was because she kept forgetting. And, although she could remember her husband, whom she had known many years, every time Dr. Dajer came into the room he had to be reintroduced to her!

Wondering what could possibly have happened to Mrs. Duke to make her memory fail in this manner, Dr. Dajer ordered a CAT scan of Mrs. Duke's brain. The CAT scan revealed the cause of her problem. She had experienced a very minor stroke caused by a tiny clot that temporarily blocked the artery feeding the hippocampus, an important part of the brain involved in memory. With treatment, the small clot broke up and the blood flow returned to the hippocampus before any permanent damage had occurred. Fortunately, Mrs. Duke fully recovered and was able to go home. But for a couple of days she had totally lost her ability to place new experiences into her memory, an ability that we all take for granted.

Next, consider the case of Mr. M. D., who had also suffered a stroke. In an evaluation following his stroke, Mr. M. D. was shown pictures of different things—toys, tools, kitchen objects, colors, clothing, animals, vehicles, food products, and other odds and ends—and asked to name what he saw. As Mr. M. D. zipped right through the list, he seemed to do as well as anyone might. Then, suddenly, Mr. M. D. stopped. He was looking at a picture of a peach. When the examining physician asked him what he was seeing, Mr. M. D. just shook his head and said that he didn't know. The next picture was of broccoli; he didn't know what it was either. The remaining pictures were of various fruits and vegetables; whether he was looking at cherries, lettuce, tomatoes, or cucumbers, the response was always the same—Mr. M. D. recognized none of them. If it grew in a garden or an orchard, Mr. M. D. couldn't name it.

Mr. M. D.'s stroke, occurring in the frontal lobe of the cortex of his brain and in the areas just below known as the basal ganglia, had apparently resulted in this form of memory loss.

Finally, let us examine the case of Mr. N. A., a college student. One day, Mr. N. A. entered his dorm room just in time to be pierced by a miniature fencing foil with which his roommate was practicing. The roommate actually thrust the tip of the fencing foil into Mr. N. A.'s right nostril and upward, piercing the middle of Mr. N. A.'s brain. With the precision of a surgeon's knife, the foil cut into the left side of a part of the brain known as the thalamus. As a result, Mr. N. A. lost most of his verbal memories (which are most commonly stored in the left side of the brain), while he retained memories of images. In situations like this it is common for the individual to be unable, for example, to find the word APPLE in a list of common fruits, while having no problem selecting a photograph of an apple from pictures of fruits.

Case studies like these show us that memories are stored in the brain in particular ways and in particular locations. Damage to specific areas will commonly result in particular memory impairment. In a similar way, if you damage the hard drive of a computer, particular data might be erased or become irretrievable, depending on which part of the hard drive was harmed.

There are companies that specialize in retrieving data from damaged computer hard drives. They often have surprising success in recovering much of the data from a hard drive that was burned in a fire, for example, or bashed in an accident. But imagine for a moment, that you took out the hard drive from your computer and just placed it into some sort of industrial strength 15-speed metal-pulverizing blender. Then, skipping past the buttons that said "stir" or "blend," you pressed the one that said "frenzy," and let the whole process go on for about a minute. Finally, you turn it off and notice that the molecules of your hard drive are now in a condition of fairly random scatterment, i.e., it's dust. But you're not worried, because you can take your damaged hard drive to one of those lost data recovery companies.

So, in through the front door you jaunt, and pour your dust onto the table of their top guy, and ask for your data back. He looks at the pile of dust and blinks a few times. Then he looks up at you and blinks some more. Then he says, very slowly so you will be sure to understand, "Your data are gone, and they are not coming back."

When we die, our brains eventually revert to their original state, *dust-to-dust*. The cells in which memories were stored die, break apart, cease to function. The data once contained within are lost—gone forever as surely as the data on the imagined hard drive that was pulverized. The idea that these memories could somehow turn up years later in someone else's brain, or even in "your" brain (supposing that it was possible for your consciousness to be somehow reincarnated), is *fantastic*, in the literal sense of the word, as it refers to fantasy. Try as I might, I can find nothing in science that would even hint at such a possibility. Between lives where would these memories be stored? How would they be transferred to the next living being to have them? Science offers no clue as to how it might be possible.

Now I know, just because science hasn't found any possible way doesn't necessarily mean that it couldn't happen. But it doesn't necessarily mean that it could, either. Still, I have some sympathy for the argument that it might be possible. I can see that if someone in Colonial America had suggested that it would be a wonderful thing if it were possible to hold a device in the hand that would enable him to have a real-time conversation with someone in China, the best minds of science would have happily informed the person that there was nary a hint of how such a thing could ever be possible without the intervention of divine forces. Then again, Dr. Benjamin Franklin might have stroked his chin and ventured that if lightning could be controlled, directed along a particular path, and somehow modulated...well, perhaps...just perhaps. But Dr. Franklin was in the habit of offering interesting counters to the general consensus.

Still, I shall play by the rules I have set down for this venture. If modern science, logic, or reason offer no clue whatsoever as to how a memory once stored in the brain of a person long dead could turn up in someone else who is currently alive, I will assume that it is not possible, or at least so unlikely as to not merit further consideration. As the late Carl Sagan was fond of saying, "Extraordinary claims require extraordinary evidence." In this vein I will dismiss claims of past life experiences as unbelievable.

There will be some, however, who will maintain their faith in such a possibility, commonly with the rallying cry, "anything's possible." You know, I have managed to get through over half a century of living without being arrested. But should a confluence of events ever juxtapose first myself holding a large cream pie, and second, someone who is saying to me, "but anything's possible," I am going to end up with a rap sheet.

One of the few things that I would argue was *impossible* was that "anything's possible." We live in a universe that appears to follow certain laws of physics. Even if the universe is nothing but chaos and the so-called "laws" are merely a measure of the mean average of the chaos (and if we can make predictions from them, how can they rightly be called "chaos?"), there will still be things that can happen and things that can't. I know that Captains Kirk, Picard, Janeway, Archer, et al., regularly travel to the far reaches of the galaxy at warp speeds and that their fans expect that someday such a thing will be possible, but perhaps it isn't. Perhaps it is a rule of nature that no person will ever be able to transport him or herself through warped space and seemingly surpass the speed of light. It might simply be impossible. It might very well be that no Earthling will ever visit anyplace that is more than a few light years away from our home planet. This could also well explain why no one has come to visit our world from another. Or, at least, not so you could see the pictures on CNN. If super-light speeds aren't possible, my guess is that if aliens ever do arrive, their first act will be to stagger from their ship and beg for food and water. Or, if they follow proper alien protocol, ask for a Big Mac and a Coke because they have been, as all proper aliens will, monitoring our transmissions for many years.

I know there are those who say that the aliens are already here and have been hanging around Roswell, New Mexico, or over at that Area 51 place. They have all these drawings of the skinny aliens, with the big heads and the huge, but sort of cute in a Disney way, bulging eyes. Speaking for myself, I am glad that I am a member of a species that has evolved small eyes located in deep sockets that are surrounded by bone and protected by eyelashes. I can only imagine that the most common phrase in the language of those aliens must roughly translate as, "Ow, my eye!"... But I digress.

WHO ARE YOU?

What is a friend? A single soul dwelling in two bodies.

—*Aristotle*

There must be something that makes you different from other people, or you'd be one of them. For the moment, let's continue with our assumption that the differences between people, and the existence of different conscious minds, are the results of differing physics, chemistry, biology, or something, and that each individual is a unique combination of whatever that stuff is. In other words, you are you because a certain combination of something makes you be *you* and no one else.

This, in turn, leads to a very interesting question. If some combination of stuff has led to the creation of you (and after all, you do exist), then what forbids this combination from ever reoccurring? Where is there a rule that says, "Now that you actually have gone and gotten yourself born, that's it, you're never allowed to go and do something like that again?" Obviously your combination can come up once, I mean, here you are. What forbids the same lotto numbers from being drawn a second time? Nothing that I can think of, other than it might simply be unlikely.

Understand that in this instance, the combination that created you would be a highly personal one. It would be you, and only you. I am not talking about an identical twin or a clone of you. I am talking about a possible reoccurrence of whatever combination of things led to *your personal consciousness*, and that such a combination might conceivably recreate your conscious awareness in a new life.

If such a thing were possible, you would be born once again. You would start life with a clean slate. No memories from a past existence could possibly be part of your new life for reasons that we have already discussed. But if this person, this new you, didn't have any of your memories, or your desires or thoughts, it seems reasonable to ask would it really be *you*?

I would argue that if such a reoccurrence were possible, that the person born would actually be you, in the sense that if that person looked at a pretty sunset, it would be you who experienced the beauty. If that person pricked a finger on a thorn, it would be you who felt the pain. If that person heard a beautiful song, it would be you who enjoyed the sound. If you had a very extensive case of amnesia, you would still be you, even though your past memories were erased. In fact, you are you now, and the possibility that you might have lived before has no bearing on that fact, so if you had a new life, you would feel as totally at home being "you" once again as you feel being you now.

The idea that our consciousness might reoccur through the chance combination of certain stuff shouldn't strike us as that odd of a supposition. The really impossible one to believe is that you or I would ever appear at all. That's the one I would have a hard time believing if it weren't for the fact that we are here now, and alive. Once we have established the amazing fact that *we are possible*, it doesn't seem half as wild an idea to imagine that we are possible a second time, a third time, or even more often.

If this is what's going on, and I am not yet suggesting that it is, then there is no death as we have come to think of death. There would only be life after life (assuming that your combination kept coming up). After all, one can never experience death; dying, yes—death, no. The instant we die (assuming that brain dead is your definition of death), we cease to experience anything. If our combination does keep coming up, then our last instance of experience in this life is followed straight away by a rebirth and a continuance of life. Death as an experience is beyond our reach. Life then might be an eternal dance, and what we call death is a brief moment during which we change partners and keep dancing. We discard one body and all the memories it contains, take on a new one, and continue.

This really doesn't sound like such a bad deal. Death as an end to everything is a pretty scary concept. But what if death is really just a fresh start? Of course, it might seem that it would be sad to leave your friends and family, but you would make new friends and belong to a new family, and gratefully, without memories of your old friends and family to cause you sadness at their absence. Yes, you would lose all of the wisdom and knowledge that you gained over the years, but you'd be trading that for the joy of experiencing things for the first time and the thrill of discovery. Still, questions remain. What really is consciousness, and can it be recreated? What are the chances of such a thing happening? And, what should we do about that nasty supervenience paradox? Perhaps we might also ask, "Into what sort of life are you likely to be reborn?"

THE GHOST IN THE MACHINE

As I was walking up the stair
I met a man who wasn't there.
He wasn't there again today.
I wish, I wish he'd stay away.

—Hughes Mearns

What is the conscious mind and what causes it to be? Discovering the answer to this one question encompasses what many consider to be the Holy Grail of psychology. I wish we had the answer, but as of now no one does. However, many have addressed the issue and we might take a good stab at uncovering what consciousness might be, since we are considering the possibility of its recreation.

In the 17th century, the great philosopher and mathematician René Descartes proposed that the conscious mind and the body were two totally different entities. The body, he argued, functioned like a machine according to the various laws of nature. Within this machine was a consciousness that functioned in parallel with the body, but was independent of it. In essence, Descartes was trying to imagine the body as a vessel for an independent soul.

Descartes went on to suggest that the only thing of which he could be certain was that his conscious mind existed. He conceded that there was no way for him to prove that his body was real. To most people it would seem that it should be the other way around. The body is real, but the soul is ethereal, and for many, doubtful. But Descartes stood this traditional view on its head with his famous observation, *"Cogito, ergo sum"* (I think, therefore I am).

What Descartes meant by this, of course, was that he had to be conscious in order to think, that is he had to exist in some form if he were thinking. But whether everything that he sensed, including the existence of his own body, was real, or just an illusion, was something that he could never know for certain. All he could say for sure was that if he was being tricked, and what he sensed wasn't real, he had to *exist* in order to be tricked! So, curiously, his conscious awareness was all of which he could be certain. The rest was pure supposition.

The general idea that a spirit might reside within the body was not new when Descartes elaborated upon it. In fact, it was quite a popular concept among many cultures, including the ancient Greeks. The Greek word *psyche*, which means "mind," denoted a spiritual presence as well as a physical one. Descartes, however, drew a clear distinction between the mind and body, arguing that while the two worked in parallel, they were not made of the same stuff. While the body might act like a machine, the mind or spirit was free of such mechanical constraints. This view became known as the "dualist hypothesis."

In 1949, Gilbert Ryle wrote *The Concept of Mind*, in which he referred to the spirit to which Descartes alluded as "the ghost in the machine." Ryle was critical of the dualist hypothesis. By 1949, neuroscience was beginning to show that different parts of the brain were directly involved in aspects of mental functioning, aspects that Descartes would have considered to be encompassed by the spiritual mind. This began to beg a reductionist view of the spirit. For example, what happens when you watch a movie? Does your body process the information through your eyes and ears and then send it through a lot of brain cells until your spirit gets to see it? Remember, to be true to Descartes' position, we can't say that the spirit is mechanical or else we'd be left asking if there were a little person inside of the big person who got to see the movie. To be true to the dualist view we have to say, along with Descartes, that the spirit is not material and not part of the body. So, then, how exactly would a ghost see and hear? It was the kind of question that modern physiologists like Ryle, who were becoming ever more familiar with the various operations of the brain, would begin to ponder.

Ryle inquired further, "What is a mental process?" If it is based on intelligence, memory, motivation, etc., why does it appear that all these aspects of the mind are located in, and controlled by, various parts of the brain (a part of the body)?

While this is in no way proves that the dualist hypothesis is incorrect, it does cause one to pause and wonder. Why is the brain, a part of the body, controlling all these aspects of mental activity that were once considered to be the province of a separate spirit or mind? Is it possible that the mind is, in fact, part of the body and that there is no separate spirit? This idea encompasses the contrasting view to Descartes' position, and is known as the "materialist hypothesis" or sometimes "monism."

The conflicting elements of the dualist and materialist hypotheses have come to be collectively known as the "the mind-body problem." The mind-body problem has not been resolved, and I am not about to attempt a solution. But I would like to offer a guess. At the start of this book, I said that what happens to us after we die is unknowable. I strongly believe that. So I set us off down the road to examine the commonly held secular view that when you're dead, you're dead, and that's that. Now, we are examining the logic behind the possibility that our conscious minds might be reborn after we die as a matter of random chance and probability. To do that it would be helpful, although not absolutely necessary, to have an idea of what the conscious mind is. But no one knows. So now, we will take a guess based on logic, reason, and empirical research.

Modern scientists generally support the materialist hypothesis. The dualist hypothesis is currently unpopular among scientists, probably because the idea of nonmaterial spirits sounds too much like ghosts and phantoms to suit the tastes of anyone engaged in "hard" science, and also because there is, so far, absolutely no respectable evidence to support the existence of any such an energy, force, or entity as a "spirit."

I want to stress that this does not mean that the dualist hypothesis is wrong. Popularity, even among an educated crowd, is hardly the same as proof, and history abounds with examples of the lone voice in the wilderness heralding a discovery that no one else believes until at some later date it is found to be true.

My problem with the dualist position is Descartes' argument that the mind is not of the body; not made of the same stuff. This would mean that the conscious mind is not the result of matter and energy, as we know these, but the result of forces or stuff hitherto unknown. Of course, I can't say for certain that such unknown forces aren't responsible for the conscious mind. As Shakespeare's Hamlet said, "There are more things in heaven and earth, Horatio, than are dreamt of in your philosophy." But not only would such a force or substance have to be unknown to us; it would have to *forever remain unknowable*! And that's the problem.

For instance, let us assume for the sake of argument that the conscious mind is created by some mysterious unknown energy or material. Such an idea is not foreign to modern physics. In fact, astronomers and physicists are currently searching for mysterious stuff and forces in the universe such as "dark matter" and "dark energy." But imagine that one day in the distant future, physicists finally came to understand the mysterious substance or force; the one that we are imagining might be responsible for consciousness. Imagine that they came to understand how it works and were then able to predict with a fairly certain degree of accuracy how it behaves. This is very much in keeping with the history of physics, inasmuch as initially mysterious particles such as the neutrino, or inexplicable forces such as the strong force, were postulated to exist, were then discovered, and finally understood well enough to be added to the physics textbooks.

If such an understanding were to be gained concerning the force or stuff responsible for the conscious mind, the once mysterious force or stuff would become just a further extension of our understanding of physics. One might then overhear comments like, "Oh, the conscious mind? Of course, it is a function of the interaction of chronons as they exchange M negative energy packets, within an 890 angstrom radius." If it ever comes to that, wouldn't the conscious mind become just another functioning and understood *part of the body?* If it is made of stuff and forces that we can eventually come to understand *and predict*, then the original conception of it as a non-material, separately functioning entity, becomes no more than an expression of our former ignorance of physics. To remain forever "not of the body" requires that we can never understand it or bring it within the purview of our empirically derived and predictive sciences—*never.*

This whole issue reminds me very much of the arguments concerning the concept of free will. The arguments favoring the existence of "free will" seem reasonable at first, but you are soon left wondering such things as, "Are you really free to make the choices you make, or is that just an illusion and the truth is that you are no freer to make a choice than a tree is free to stand or fall?" It is an interesting question and the subject I dare say of a different book. But allow me to just note that if you ever came to understand the biophysics of whatever "free will" is, you could then ask such questions as, "How much free will do you have?" or "Can you in-crease or decrease your free will?" Questions like these paradoxically decry the existence of such a thing as "free" will. If it is predictable, it can't be free, anymore than a cannonball is free to follow its own path.

Similarly, if you could ever come to understand and predict the me-chanics of the conscious mind, no matter what undreamt of energies or stuff might be involved, it would then be properly classified as part (albeit a very exceptional and interesting part) of the body. This is why I argue that for the dualist position to be correct, the conscious mind must not only be distinct from the body, it must remain ever unknowable to our science.

Of course, who knows, that might be how it is. But I am hard pressed to imagine how any functioning entity might behave in a way that was wholly unpredictable, inasmuch as our science could never come to understand it well enough to make some predictions about how it would act. Then again, predictability implies that some sort of rules of action are present, a lawful set of principles by which actions occur (in the manner of the law of gravity). But to be a separate entity from the body, the dualist version of the mind cannot obey such mechanics lest it become part of the body. Would the dualist "mind" then be random in its actions? If so, that still doesn't help. Randomness is highly predictable (although not absolutely predictable). If it weren't, no casino operator could hope to make a living. Truly, if consciousness is not of the body, not material, then its manner of functioning becomes literally inconceivable.

How then might the conscious mind be part of the body? Perhaps modern brain research can give us a clue about how consciousness could arise within the brain itself without the need for a spirit's presence.

PETAFLOPS AND DATA CRUNCHING

Do Androids Dream of Electric Sheep?
　　　　　　　　　　　　　　　　—Philip K. Dick

You are probably familiar with the prefixes *mega* and *giga*, as in "megabyte" or "gigahertz." Mega refers to a million and giga to a billion. A 500-gigabyte hard drive, therefore, can store 500 billion bytes of data.* The next prefix beyond that is *tera*, for a trillion. After that, comes *peta*, which refers to a quadrillion, or if you prefer, a million billion.

Computers generally run calculations based on "floating points." IBM defines a floating point as "a method of encoding real numbers within the limits of finite precision available on computers." However, we needn't concern ourselves with anything more than the fact that the raw processing power of a computer is often figured by how many floating-point operations per second, or FLOPS, it is capable of producing.

IBM is hoping that its most powerful computer, *Blue Gene*, which is dedicated to operations involving genetics, might be capable of an amazing one petaflop by the year 2012. A petaflop computer could perform one quadrillion calculations per second! Truly, that is an awesome computational power.

But here is an interesting thing to consider. It is, in fact, possible to create a portable computer (that would weigh less than 4 pounds), that would not only operate at a rate of an astounding 5 petaflops, but which would do so while only requiring 70 millivolts of electricity (meaning that you could run 12 of them on a single 9 volt battery)! Let me also add that

---

* A byte is usually defined as a string of binary code (0s and 1s) that is 8 digits, or "8 bits," long).

this stunning computer wouldn't overheat because it would run at temperatures of less than 100 degrees Fahrenheit--easily cool enough to touch.

Computer engineers have known for a long time that a computer with such processing power was physically possible. They know this because we all possess one. It rests between our ears. Computers can run specific calculations faster, but in terms of sheer number of calculations per second, there isn't yet a computer that can match the human brain. Perhaps in 10 years there will be, but not yet. Amazing as it might seem, that little 3½ pound body organ in your head is crunching data at a rate of about 5 petaflops!

We take our brains for granted so much that it sometimes comes as a shock to consider just what amazing organs they are. I am reminded of a neurosurgeon I once knew who would lecture to his residents about various surgical techniques, and often end his lectures with a reassuring, "don't worry, after all, it's not brain surgery." Then he would pause, reflect, and say, "Well, okay, I guess it is brain surgery." I am reminded of him because of an incident in which he was suddenly taken aback by what it was he was actually doing when he operated. Allow me to tell you the story.

In order to conduct brain surgery, especially if it is necessary to cut far into the brain to, let's say, remove a deep seated tumor or repair a deep underlying blood vessel, the surgeon has to be very careful not to destroy a vital area as he or she cuts down through the patient's brain to get at the problem. To do this the surgeon often enlists help from the patient.

The brain itself is relatively impervious to pain, so after giving a local anesthetic that allows the surgeon to cut open the patient's scalp and remove a portion of the skull, the brain can be exposed while the patient is awake. The surgeon can then stimulate different parts of the brain with an electrode and ask the patient what it is that he or she is experiencing.

Although brains are very much alike in many respects, with certain areas responsible in part for vision, hearing, touch, emotion, etc., each person is an individual and you can never be entirely sure how close you might be to cutting a vital area without checking. During this procedure, the surgeon might touch an electrode to a certain part of the brain, and the patient might report a tingling on the left side of his lip. That lets the surgeon know that he is probing an area responsible for receiving touch sensations from the lip. Or if the lip quivers, he might realize that he is stimulating an area of the brain responsible for moving that part of the lip. If he had to cut through either of those portions of the brain, his patient might end up with a lip that was either numb to feeling, or paralyzed, perhaps a sacrifice worth making if it saves the patient's life. If, however, the electrode is placed and the patient says, "It feels funny," but says it with distinctly slurred speech, the surgeon will know not to cut there, for if he did, he might damage his patient's ability to speak. The surgeon will then look for a better avenue down to the problem, unless there is absolutely no other way in.

The surgeon who brought this story to mind was telling me one day of an experience he had while mapping a patient's brain using the technique just described. It was then that this particular surgeon was unexpectedly overcome by a sense of awe. He told me, "You know, I had done that procedure at least a hundred times, but it just suddenly occurred to me that I was talking to a body organ and it was answering back! Really," he said, "you can talk to a heart or a liver all day long and you might just as well be talking to a rock. But here I was, having a conversation with this body organ about how it felt when I touched it, and it was telling me! For a couple of seconds I was a little freaked, and then I got back to business."

Our modern conception of the brain is, in fact, so stunning that there is little reason not to be "freaked" by it. Interestingly, the ancients often had different ideas about the purpose and function of the brain. Aristotle, the great Greek philosopher, believed that thought resided in the heart, and not the brain. He based this on the observation that so many powerful feelings were felt in the chest, while the brain, with its many convolutions and folds, was more suited to cooling the blood, something akin to a radiator. But people who have died and had their hearts transplanted to another, don't seem to wake up in a new body. And when activity in the brain, especially in the cerebral cortex diminishes owing to a good dose of anesthesia, we experience a loss of consciousness. It would seem that, somewhere in that amazingly complex organ, is the seat of consciousness.

Where might consciousness be? Where in the many folds and crevasses of the human brain might it reside? Perhaps we might take a moment to consider some important aspects of the human brain.

Is consciousness to be found in a particular place in the brain? Perhaps—but we must remember that the brain is a living organ, highly adaptive and flexible. The classic case of the Silver Spring Monkeys is a good illustration of this point. In the late 1970s, a neuroscientist named Ed Taub, working at an institute in Silver Springs, Maryland, began a study on some monkeys in which he destroyed the point where the sensory nerves from one arm entered the spinal column. Taub was going to use the monkeys in rehabilitation experiments, but before he was able to do so, the monkeys were confiscated following a raid on his laboratory by animal rights advocates. It was only through the efforts of the animal rights advocates that the monkeys were kept alive for years beyond the time that they would normally have been destroyed. Ironically, this led to an amazing discovery.

Since the operations to sever the nerves of their arms from their spinal columns, these monkeys had "remapped" their cerebral cortexes of their brain. Points on the brain that once reacted to a touch on the arm now reacted to a touch on the face. It appears, then, that when we learn something new or compensate for an injury, not only do we make new neural connections during the process but also the brain often adjusts itself so as to dedicate a larger portion, or even a new portion, of itself to the demand.

Another good example of the adaptive flexibility of the human brain comes from modern brain scans that have shown that when an adult who is blind learns to read Braille by running a forefinger over the raised markings on the paper, markings that stand for symbols, a location on the surface of the brain where incoming sensory messages from that particular fingertip are processed, will begin to expand into areas previously devoted to *other* fingers. In other words, the brain is responding to the demand to learn Braille by giving over more processing space to the fingertip involved.

Sometimes, many differing aspects of the brain will work together to form a whole picture or concept. Perhaps one of the best examples of this ability of the brain, and how such ability can be lost, comes from the writings of neurologist Oliver Sacks. In his book, *The Man Who Mistook His Wife for a Hat*, Sacks spoke of a music teacher, Dr. P., who had come to see him in the company of his wife, because Dr. P.'s wife had said that he was having severe problems. On examination, however, Dr. P. seemed all right. He was charming, friendly, and intelligent and appeared quite normal.

When the examination was completed, Sacks noticed that his patient hadn't put his socks back on. Then, Dr. P., indicating his foot, asked, "That is my shoe, yes?" Sacks, somewhat in a state of shock, said, "No, it is not. That is your foot. There is your shoe." Dr. P. responded, "Ah, I thought it was my foot." Obviously, the examination was not finished! What in the world could be wrong?

Sacks then showed his patient a picture from a magazine and asked him to tell what he saw. Dr. P.'s eyes darted about, picking up bits and pieces of the scene, but failing to get much out of it. Acting as though he had done well on this test, Dr. P. prepared to leave; then he reached over and took hold of the top of his wife's head and tried to lift it on top of his own. He had mistaken his wife for his hat! During a later examination,

Sacks tried a test that's worth describing because it highlights Dr. P.'s problem. In Sacks' own words:

> "What is this?" I asked, holding up a glove.
>
> "May I examine it?" he asked, and, taking it from me, he proceeded to examine it...
>
> "A continuous surface," he announced at last, "infolded on itself. It appears to have"—he hesitated—"five outpouchings, if this is the word."
>
> "Yes," I said cautiously. "You have given me a description. Now tell me what it is."
>
> "A container of some sort?"
>
> "Yes," I said, "and what would it contain?"
>
> "It would contain its contents!" said Dr. P. with a laugh. "There are many possibilities. It could be a change purse, for example, for coins of five sizes. It could..."
>
> I interrupted the barmy flow. "Does it not look familiar? Do you think it might contain, might fit, a part of the body?"
>
> No light of recognition dawned on his face. No child would have the power to see and speak of "a continuous surface...infolded on itself," but any child, any infant, would immediately know a glove as a glove, see it as familiar, as going with a hand. Dr. P. didn't.

Dr. P. as it turned out, had an inoperable tumor in his brain. It had caused his strange problem. Dr. P. could see in the way that a computer might be able to see. He saw lines, angles, details, shapes, but he never could realize the "big picture;" he couldn't see the "gestalt;" he couldn't see "what" it was he was seeing. He knew the shapes and angles of his hat (and his wife's head), but he couldn't comprehend the wholeness of "wife's head" or "hat." This lost understanding was apparently something that would have been provided by the right hemisphere of his brain, had it not been damaged.

Even with this strange problem, Dr. P. was able to function with the help of his wife, who laid out his clothes each day in exactly the same spot and arranged his breakfast and other routines in a set order. In his way, Dr. P. knew he was, in fact, putting on clothes and eating breakfast. He was even able to teach music, which he did until his death brought on by the tumor some months later.

Perhaps consciousness, then, might best be understood in terms of awareness, as opposed to what sorts of things we might do that are unconscious. But before we draw any conclusions, let's take a look at one more area of brain research.

The human brain is divided into two hemispheres, a left and a right (see Figure 2). Occasionally, when a person suffers from a severe form of epilepsy and cannot be treated in any other way, a surgeon will sever the few connections that exist between the two hemispheres of the brain. This prevents the epileptic seizures from spreading from one hemisphere to the other.

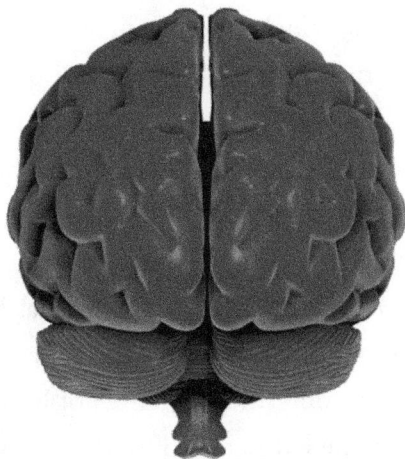

Figure 2. The human brain is divided into two hemispheres, a left and a right.

In the early 1950s, R. E. Myers and R. W. Sperry made a startling discovery when examining cats that had their two hemispheres surgically separated. They found that the right hemisphere could learn something while the left hemisphere remained ignorant of what had been learned, and vice versa. Over the next three decades, Sperry, working at the California Institute of Technology, and Michael Gazzaniga, working at Cornell University Medical College, added to our knowledge about the divided hemispheres. In sum, in human patients with split brains, the two hemispheres of the brain function separately; their interconnectedness has been disrupted.

If you were to meet a person with a split brain, you most likely would not notice anything unusual. To find a difference, you would have to conduct some interesting tests—for instance, tests of the seeing process. Figure 3 shows how the nerves that carry information from the eyes are organized. Each eye independently sends messages to *both* hemispheres. The visual system has evolved so that the left halves of *both* retinas (the retina is the back part of each eye, which contains light-sensitive neurons, or nerve cells) send their visual messages to the left hemisphere, whereas the right halves of *both* retinas send their messages to the right hemisphere.

Even split-brain patients looking at something with just one eye will receive visual messages in both hemispheres. But, if the patient glimpses something briefly out of the corner of his eyes the situation is different. Look again at Figure 3. Imagine that something has been briefly glimpsed in the far-right field of vision while your eyes were looking straight ahead. Due to the extreme angle, only the left halves of both retinas will respond to the object. Similarly, the left halves of both retinas cannot sense objects far off in the left field of vision. Consequently, when both you and a split-brain patient glimpse something out of the corner of your eyes, the object will be projected to only one hemisphere. Unlike you, however, the split-brain person's receiving hemisphere cannot share this information with the other hemisphere, because the connections between the two hemispheres have been cut.

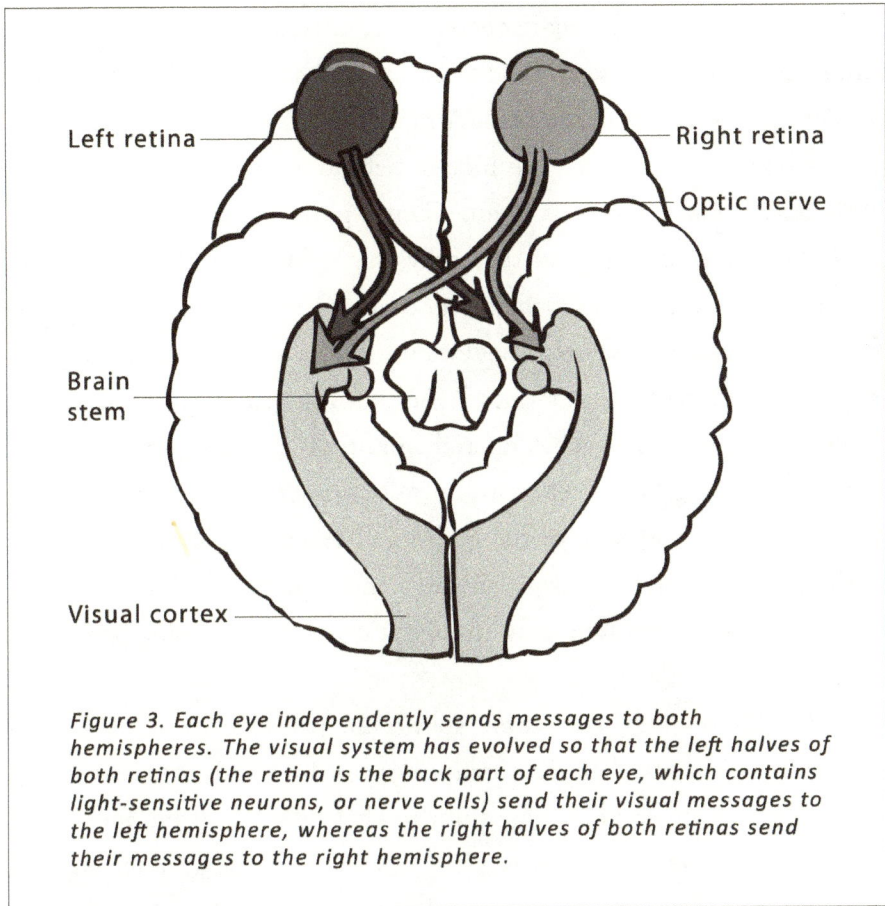

Figure 3. Each eye independently sends messages to both hemispheres. The visual system has evolved so that the left halves of both retinas (the retina is the back part of each eye, which contains light-sensitive neurons, or nerve cells) send their visual messages to the left hemisphere, whereas the right halves of both retinas send their messages to the right hemisphere.

A split-brain person usually has no trouble talking about such visual messages when they are sent to the *left* hemisphere, because that is where the language capable portions of the brain are typically found. Whenever an object or photograph is presented to him, he can tell you what it is. However, if a visual image is presented off in the *left* visual field, so that it is sent only to the *right* hemisphere, the split-brain person might be unable to tell you what he has seen. This is because the right hemisphere typically has poor language and speech ability compared with the left hemisphere. In most people, the left hemisphere can speak and describe what it has seen, but the right hemisphere, although it too can see, often cannot say what it has seen. But it has, in fact, seen the object. As proof, consider the following experiment. A number of objects, such as a spoon, a block, a ball, and a toy car, are hidden behind a cloth barrier so that the split-brain patient cannot see them. Then a photograph of one of the objects is shown at an angle, so that only the right hemisphere registers it, while the left hemisphere remains ignorant of what has been seen. The split-brain patient will say that he does not know what the picture was (after all, his left brain doesn't know). But, although the split-brain patient usually cannot articulate what his right brain saw, he can find the correct object by reaching behind the curtain *with his left hand* (each hemisphere controls the opposing side of the body) and feeling for it. This is because the right brain, which controls the left hand, knows what the object was and what it should feel like, even though it cannot put this knowledge into words!

It should be noted that some split-brain patients do not show such remarkable differences between hemispheres, which indicates that in some people the two hemispheres might not be that different from one another. How different the hemispheres are from each other, is referred to by neurologists as "lateralization." People who are highly lateralized have hemispheres that are markedly different. Interestingly, women tend to be less lateralized than men and are less likely to lose language ability following a left hemisphere stroke. However, lateralization also varies considerably from one person to the next as well as across genders. This illustrates how variable brain organization can be from one person to another. Generally, however, it does appear that the left hemisphere of the brain is organized in a way that facilitates language recognition and production in most people.

Split-brain research has also yielded information that indicates the existence of a very special "interpreter" in the left hemisphere. To illustrate, let's look at an experiment conducted some years ago by Michael Gazzaniga on a 15-year-old boy with a split-brain named P. S. In a laboratory, two pictures were shown to P. S., one to each hemisphere. The left hemisphere was shown a chicken claw, whereas the right saw a snow scene. P. S. was then shown a number of pictures on cards and asked to pick the one that went with the picture he had seen. His right hand pointed to a picture of a chicken (remember, his left hemisphere had seen the chicken claw), and his left hand pointed to a snow shovel (his right hemisphere had seen the snow scene—see Figure 4). This is what Gazzaniga had expected. But, when P. S. was asked why he had pointed to these objects, he answered right away, "Oh, that's easy. The chicken claw goes with the chicken, and you need a shovel to clean out the chicken shed."

Stop for a moment and consider what this means. The left hemisphere had no idea why the left hand had pointed to the shovel, so it fished around for a logical reason and came up with one—and was wrong. Based on this work and many other examples gathered over the years, Gazzaniga has come to believe that there is an "interpreter" in the left hemisphere that is independent of the language areas that "constructs theories about...actions and feelings and tries to bring order and unity to our conscious lives. [It] appears to be unique to the human brain and related to the singular capacity of the brain to make causal inferences."

All of the various areas of the brain interact with one another to process information, so it should come as no surprise that the brain would try to make sense of information gathered from different locations within itself. The "interpreter" that Gazzaniga discovered appears to be one example of such an effort.

Figure 4. In a laboratory, two pictures were shown to P. S., one to each hemisphere. The left hemisphere was shown a chicken claw, whereas the right saw a snow scene. P. S. was then shown a number of pictures on cards and asked to pick the one that went with the picture he had seen. His right hand pointed to a picture of a chicken (remember, his left hemisphere had seen the chicken claw), and his left hand pointed to a snow shovel (his right hemisphere had seen the snow scene).

The way in which the brain attempts to process all this disparate information is interesting. Researchers have discovered that the brain organizes such efforts according to three basic principles that are worth taking a moment to consider if we wish to pursue an understanding of what consciousness might be.

The first principle states that the brain is *interconnected*. This refers to the fact that many different parts of the brain will spring into action when confronted with a single task, something we have just discussed in the forgoing example. The brain appears to operate via interconnection "modules" that are relatively independent. A brain module is simply a grouping of cells appearing in a strip of brain tissue perhaps no thicker than 1/10,000th of an inch. Each module of the brain has a task to do and will work in *parallel* with other brain modules that are working on other aspects of the same task. Once each module has done its job, the efforts of the individual modules are combined and *integrated*, which is what is meant by the "interconnectedness" of the brain. Evidence for this principle of brain organization comes from stroke victims and others with some form of brain damage that has interfered with the brain's interconnections. In these individuals certain integrations do not occur. For instance, Michael Gazzaniga has reported a patient who, after some very specific damage, was able to name fruits *or* red objects (each task assumedly controlled by a different module) but not red fruits (a task probably requiring the integration of two modules working in parallel). Gazzaniga also described a patient who was able to compare two words, but only if the words looked or sounded alike. Different sounding words required integration for comparison—one that this individual was unable to perform.

The second principle states that the brain's organization is generally *hierarchical*. Information coming into the brain from the senses, or perhaps even that gathered from the memory, appears to be built up from small, simple units into a comprehensive whole. For example, cells in the eye might be responsive to light, dark, or certain colors. As information from the eye enters the brain, more complex cells receive this input, cells that are responsive to lines and edges *as well as* to color and brightness. This information, in turn, is passed on to more complex cells that are additionally responsive to motion or direction of movement. This information continues to build up into more complex collections in a hierarchical fashion until whole modules finally handle it, because no single cell or small group of cells can deal with the growing complexity. Some parts of the brain might not be organized in a fashion that is all that hierarchical, and those parts might function parallel to one another. It is an area of hot debate among brain researchers.

The third principle states that there is *functional differentiation* in the brain. This means that certain brain areas are responsible for certain brain functions. For instance, certain areas of the brain are mainly involved with hearing, whereas other areas are mainly involved with vision. Even in these instances, however, there might be much overlap.

It is still very difficult to understand just how functionally differentiated the brain might be. Part of the problem is that whole areas of the brain are still a total mystery; no one knows what they do.

THE GENERATION OF CONSCIOUSNESS

*I believe that the most central fact about my existence is I perceive
that there is an "I" that observes the world from someplace inside my head.
It makes no difference how many details you tell me about the
working of the brain and the firing of my neurons.
Until you have explained how I come to that central conclusion about
my own existence, you have not solved the problem of consciousness.*
—James Trefil

But where within the brain, within all of these specialized areas, is the seat of consciousness? Does it lie in areas of the brain yet unknown? Is your consciousness the result of a brain module that rests somewhere in your cortex? If it were, would it be possible someday to remove that module and place it in a different body? If that were done, would you wake up after the surgery to find yourself now residing in that other body?

Most brain researchers have come to doubt that consciousness is a function of a highly specific brain module, or even that it might rest within the confines of a particular brain location. But if it is not a certain part of the brain, or in a certain location, where or what is it?

Let's look over some of the brain research and case studies we have been discussing while we consider what we mean by consciousness. When I think of myself as conscious, I mean that I am aware. I sense what is happening around me and I think about it. I might recall something from memory and think about that as well. In all cases, I feel like a central player dealing with sensory information, emotions, memories, and thoughts. There is a sense that "I", whomever that might be, is aware of all of these things and is living somewhere inside my head.

Dr. P., who mistook his wife for his hat, was aware of his world, but there were some things he could no longer understand in their wholeness. To me, it doesn't seem that he lacked consciousness; rather that he had a failure of intellect, or perhaps a failure of sensory integration. I can't be inside his head, so I can't know. But his condition didn't appear to eliminate his conscious mind, any more than did the conditions of patients we discussed who had memory failures due to various forms of brain damage. They were conscious, but made errors.

The split-brain patient discussed by Michael Gazzaniga might be a different case, however. In that instance, his left hand chose the snow shovel after his right brain saw the snow scene. However, "he" seemed unaware of that fact, even though his left hand was unaware. As you recall, his conscious mind appeared to have no idea why his left hand had chosen the shovel. Or am I wrong in that assumption? Perhaps his conscious mind was aware, but because his right hemisphere had little or no language capability, he was simply unable to *say to himself* what he had been aware of concerning the snow scene. This also brings to question how much language might play a role in consciousness.

At first glance, it seems that I am conscious because I am aware of what I see, hear, smell, taste, or touch. However, if I close my eyes so that I can no longer see, I remain conscious. If I plug my ears so I can no longer hear, and then perhaps float suspended in a lukewarm immersive bath until my sense of touch nearly shuts down, I remain conscious. Studies like that have been done and are sometimes called sensory deprivation studies. Most people who engage in them have found the experience very relaxing, but clearly reported that, even without much sensory input, they continued to think about things and to "see" things in their mind's eye. Consciousness, therefore, is not dependent upon continuous sensory input from external sources.

Is it thinking then, and awareness of thought, that defines consciousness? Perhaps awareness is the key concept. If somehow you had no sensations, and were unable to think, perhaps consciousness would cease, if for no other reason than you no longer had anything about which you could be aware.

Is awareness a brain function? Can we look at the three great principles of interconnectedness, hierarchical structure, and functionality, and come to the conclusion that awareness is a higher function performed by some specialized part of the brain enabled by its connections with other functions in the brain? If that is true, then where am "I" in my brain? Where are "you" in your brain?

It appears that we might be coming back full circle to once again look for the elusive ghost in the machine—to look for the spirit inside who watches the images the eyes send to the brain. Is that little "person" in there? Perhaps we could trace a visual image as it makes its way into the brain—and thereby discover the final "viewer." Such searches have been conducted. Let's try it and see what happens.

Imagine a person looking at a photograph of a very pretty red rose. Also printed on the photo is the word ROSE. The image of the flower and the printed word pass through the lenses of the eyes and project upside-down onto the retinas (see Figure 5 for an example).

The retinas may be properly thought of as the first parts of the brain to be exposed to the image. The first layer of the retinas to be exposed to the image is filled with photoreceptor neurons. These cells are sensitive to light. When they are stimulated by light, their neural messages travel through a second layer of the retina that further modifies and processes the image. The last layer of retina contains what are called ganglion cells.

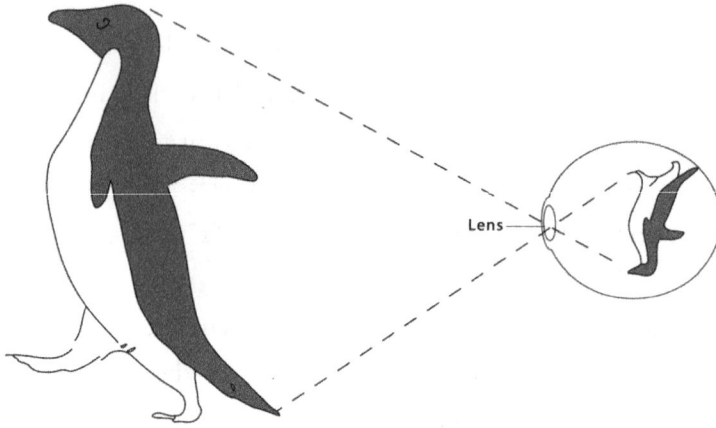

Figure 5. Any image that passes through the lenses of the eyes will be projected upside-down on the retina.

There are over 20 kinds of specialized ganglion cells. Each kind represents a complex pattern of light created by light-sensitive neurons at a particular location on the retina. Each ganglion cell, then, responds to what is called a receptive field. For example, one kind of ganglion cell may respond most to a receptive field comprising a red spot on a green background. Another kind of ganglion cell may respond most to a green spot on a red background. Still another may respond to anything that moves in a certain direction, and so on.

In this final layer of the retina, then, cells specialized to detect certain kinds of complex stimuli are already in place. From the retina, the visual information travels up the optic nerve of each eye and into the brain. But what happens to the message next? What we observe is that the visual message passes through increasingly specialized detectors (cells that only respond to unique patterns, forms, or colors), each, it is assumed, with its own receptive field.

Of course, this view of how the brain makes sense of vision leaves us with a distinct problem. There have to be limits on how specialized visual detectors can become, simply because at some point too many cells would be required to cover all the possible combinations. We can't expect that we will eventually discover individual cells only responsive to receptive field inputs for lavender SUVs traveling left-to-right at a distance of 5 meters. Even very complex cells, then, could not hope to recognize an entire image but would need to share information with millions, perhaps billions, of other complex cells.

Up to this point, we can clearly see the three main principles of brain operation at work. The visual information is *integrated* as cells combine and share data. The process is *hierarchical*, as information proceeds to ever more complex cells. And, we see *functional differentiation*, as visual information collects primarily where cells receptive to visual data appear to concentrate. That's all well and good. But now, where does the information that our subject is looking at, a red flower and a caption that says ROSE, eventually go? Where is the movie screen in the brain on which it shines, and where is the little person who watches it? Where, in all those cells, is "she" or "he?"

If you follow the visual message created by having looked at the rose and its caption, you can see it going deeper and deeper into the brain, and stimulating more and more cells, and you can even trace it to where it concentrates in the occipital lobes at the back of the brain. But you will not find some major terminus there; there is no "Grand Central Station" where all the visual messages come to a stop. Instead the messages continue on, looping back into other parts of the brain and becoming less concentrated.

Areas associated with reading printed words become active, probably as a result of the presence of the caption. Memory areas become involved and areas associated with the sense of smell as well, perhaps because a scent has been associated in the past with a visual sensation of a rose.

The information also disperses in hard to track ways, as a single neuron receiving an input might well stimulate, and send messages on to, hundreds or even thousands of other neurons. Eventually the dispersion seems to settle down as most of the neurons stimulated actually become "hyperpolarized" and therefore become harder to fire. Were it not the case that most neurons, in fact, acted like breaks, then the slightest stimulation would send the entire brain into a great convulsion. This is, in fact, more or less what happens during an epileptic seizure: Too many cells fire that shouldn't, and the brain becomes overwhelmed.

But still, after watching all this activity caused by sensing a picture of a flower and caption, watching neurons flash messages all about until things finally quiet down, we still have no answer to the seemingly simple question, "Where is the viewer? Where is the consciousness that sees the pretty blossom and reads the name of the flower?"

I suppose it is possible that somewhere in the brain there is a module that is the seat of consciousness, and that all memory, thought, or sensory stimulation sooner or later stops there to let "you" know what's going on, but I doubt it. If it were true, we might end up with a dilemma remarkably like the one that arises when we try to imagine a spirit that is not of the body. If such a module existed, what would be inside of it? Wouldn't it just be a mass of neurons and other brain tissue, much like the rest of the brain? Wouldn't we then just look at all these cells and wonder anew, "Where is the 'you' in the little brain that is inside of the big brain?" This dilemma has led many modern researchers to believe that consciousness doesn't exactly exist in a certain place or module, but rather is *generated* by the various parts of the brain as they go about their business.

This is not a new idea, and can rightly be first attributed to the great 19ᵗʰ Century British philosopher, John Stuart Mill, who argued for the concept of *mental chemistry*, as he so aptly named it.

Ironically, I have chosen a simple example from physics to demonstrate what Mill meant by mental chemistry, and I trust that chemists won't mind. Imagine for a moment that you have a disk about 12" in diameter. The disk has a small hole in the center, not unlike a CD, so that the disk can be placed on a spindle. The spindle is attached to an electric motor so that the disk can be stood up and made to spin rapidly while you look straight on at it (see Figure 6).

The disk itself is divided into three equal portions, in pizza pie style, with one-third being red, one-third green, and one-third blue. These segments of the disk are brightly colored; the colors are rich and saturated. We're talking fire engine red here, as well as emerald green, and royal blue. It is a very pretty disk if I say so myself.

Now, place this tri-colored disk onto the spindle and turn on the motor so that the disk spins very rapidly. What will you see when you look at the colored face of the spinning disk? Perhaps you are familiar with this experiment in psychophysics. But even if you are, it is still amazing to watch what happens. The bright tri-colored disk turns a brilliant snow-white. I mean—the disk will appear to be a solid and very bright white! There is not a hint of any color. And yet, should you bring the spinning disk to a stop, the three rich colors reappear and there is not the slightest hint of white, anywhere. If you could go back in time and put on such a show for King Arthur's Court, it's the sort of thing that would have made Merlin very jealous. Of course, finding an outlet in which to plug your electric motor would have been a problem....but I digress.

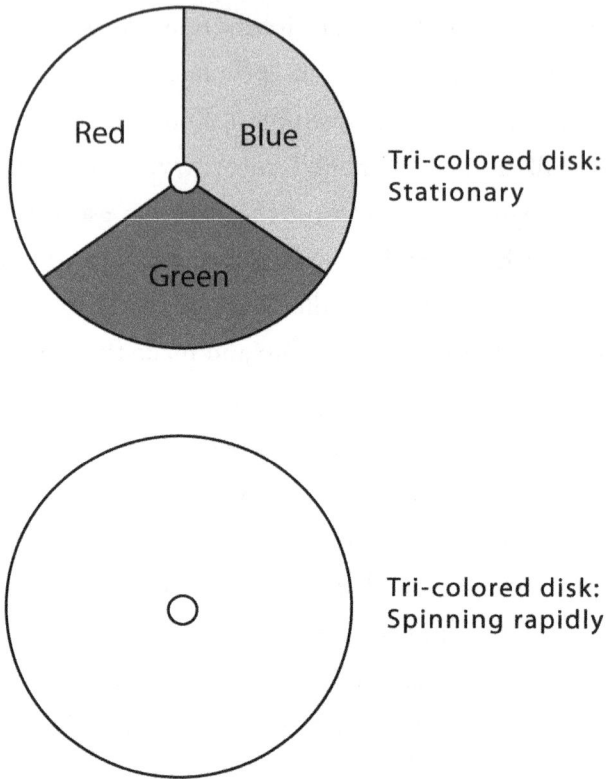

Figure 6. When a disk composed of equal proportions of the three primary colors red, green, and blue, is spun rapidly, the disk will appear to be white. White, in this instance, does not "exist" until it is generated in the mind of the observer.

By mental chemistry, John Stuart Mill was referring to how the human brain was appearing to mimic chemical processes, processes that would often allow for the combinations of interesting elements or molecules to create what seemed to be something wholly new. Here is an example: Take chlorine, a green and poisonous gas, and combine it with sodium, a volatile metal (which, in its pure state will burst into flames if exposed to air), and the result will be common table salt—seemingly unrelated in any reasonable way to the original volatile starting materials.

Early in the 20ᵗʰ Century, German psychologists dealt with this issue by formulating what became known as Gestalt psychology. *Gestalt* in this sense refers to "wholeness." The Gestalt view grew, in part, out of the desire to explain how a series of still images shown in rapid sequence could result in the perception of motion, when there was, in fact, none. In other words, these psychologists were attempting to understand how a motion picture worked, psychologically. We call them "movies" because what we see on the screen appears to move, but in fact, every image we actually see is unmoving.

The conclusion reached by the Gestaltists was, basically, that no explanation was required. It is simply a real phenomenon in its own right, and any attempt to reduce it to something simpler was pointless. At first glance this argument might not seem revolutionary, but it was. Prior to this, psychologists had been attempting to break experiences down into basic or fundamental components, assuming that by breaking experiences into elemental parts, they could be better understood. Instead, the Gestaltists were arguing that the whole experience was not just the sum of its parts, but was *more*. To a Gestalt psychologist, the spinning tri-colored disk is certainly the sum of its parts—the sum of red, green, and blue. But it is also more than the sum of its parts; it is white. If you had all of the parts of an automobile, you would have all of the parts of an automobile. However, if you assembled them, you would still have all of the parts of an automobile, but something else as well—an automobile! *The whole is greater than the sum of its parts.*

In this way, human consciousness might be *generated* by the existence of different brain modules and their interactions. Like the white color on the spinning disk, consciousness might be the sum of the interacting brain parts, but not itself a distinct part of the brain. Perhaps when your memory interacts with incoming sensations and language takes part in the form of thought, or perhaps your mind's eye, or ear, plays a role—when all these functions begin to work in unison—then perhaps the illusion of consciousness appears and we feel "self-awareness." I say "illusion" only in the sense that like a mirage, you can sense it—but if you go looking for it, you won't find it.

DO ZOMBIES WALK AMONG US?

Wandering between two worlds, one dead,
The other powerless to be born,
With nowhere yet to rest my head,
Like these, on earth I wait forlorn.
— *Matthew Arnold, "Grande Chartreuse"*

If consciousness is generated in this fashion, and it seems most reasonable to me that it is, then we immediately face some interesting prospects. Is it possible that with a little tweaking, we might keep all of the brain functioning, and yet prevent a consciousness from being generated? Is it possible that such a thing could happen on its own? The analogy that comes to mind is that of the motion picture projector. The projector is running and showing a movie onto a screen. The film is feeding through the sprockets and everything is working well. Then, suddenly, the lamp burns out. But the projector doesn't stop and the film keeps right on feeding through as though nothing at all were wrong. It's just that there is no longer a picture showing on the screen. Could the same thing happen to people?

Is it possible that through some small change in your brain—perhaps a tiny stroke like the one that caused Mrs. Duke to forget any new experience—consciousness might "blink out" while leaving the rest of the body unaffected? If something like that were possible, "you" might die, and yet no one around you would be aware of that fact. Your conscious mind would be gone, but your body would still get up in the morning and perhaps make coffee, go off to work and interact with friends and family, all quite normally; only inside your head there would no longer be anyone at home.

Or perhaps consciousness is a trait that some of us have and that others don't. As Descartes said, the only thing of which he could be certain was that *he* existed. There was no way to know if anyone else did. Sure, you can ask others, and they might say, "Well, of course I am conscious." At least that's what their bodies will tell you, since their bodies compute that whatever it is they are experiencing is what everyone else has been calling consciousness, but in fact, their front porch light is out. They would appear to be humans—active, alive, and behaving like humans, but no one would live in their bodies. I suppose another word for this is *zombie*. This whole concept is especially curious because it yields a materialist cause for what, at first glance, would appear to be a dualist phenomenon.

Perhaps zombies walk among us, perhaps they don't. Perhaps "you" can "die" years before your body actually stops going to work and interacting with others, perhaps you can't. Perhaps you might be a zombie for many years and then suddenly awaken—there is no way for you or me to ever know.

What, you say—that can't be because you remember living all those early years of your life. Well, were "you" there for those experiences, or are "you" only now tapping into the memory that your body recorded of those times? Just because you can remember it, doesn't mean that "you" were there for it. Perhaps conscious people are a minority and zombies are the majority (sometimes, especially on the highway, it does seem that way). Any of this might be possible—there is no way to ever prove otherwise.

We shall return to zombies from time to time, but meanwhile let's not worry about who is a zombie and who isn't, lest we start a whole new form of discrimination. Instead, let us consider another interesting possibility. Computers have memory. They have processing chips and many are now being given rudimentary senses. Some are also being given language capabilities. What happens when, some day, a computer is so jammed full of wonderful new processing powers that all the "spinning colors" suddenly generate "white?" Could it happen? Could silicon, operating in a way that mimicked the human brain, in fact generate a real mind? What happens when someday your computer says, "Would you please leave the light on when you leave the room? I get scared when it's dark."?

In 1950, in a paper entitled *Computing Machinery and Intelligence,* the brilliant British mathematician and World War II code-breaker, Alan Turing, argued that someday computers would rival or even surpass human intelligence. He went on to state that there might be a way to test computers to tell if they had accomplished such a feat. To achieve this end, he proposed the now famous *Turing Test.*

Turing argued that because consciousness is subjective (as Descartes noted, one can only be certain of the existence of one's own thoughts) we can never know for certain if a computer is thinking, and can only deduce that it might be thinking, or at least be intelligent, by asking it questions. The premise of the *Turing Test* is basically this: If the answers the computer gives to our questions cannot be distinguished from those given by a human, the machine has shown intelligence. As of yet, no computer has been able to pass this test while having a "conversation" with a person via a connection through a monitor that prevents the person from seeing with whom he or she is conversing.

Some have taken the possibility of a computer passing the *Turing Test* a step further and argued that by passing such a test a computer would be demonstrating actual thought; in other words, such a machine would be conscious. Others, such as the mathematician Roger Penrose, argue that such a thing could never happen because human thought requires a human brain, and that silicon based systems cannot yield the sort of physical parameters necessary for consciousness. This view, in turn, has been challenged by those who argue that such a supposition makes little sense without a full knowledge of how consciousness actually works, or what causes it to exist in the first place. Who's to say that it couldn't occur in a machine, or that the manner in which it occurs in the human mind is the *only* way that consciousness can be created?

But how would we know if a machine has become conscious? Since we can't even be certain that our fellow human beings are conscious, truly there would be no way to ever know if a machine had acquired thought or awareness. In the final analysis, we would have to take the computer's word for it!

Would we ever be willing to accept the word of a machine that it was aware and conscious? As a psychologist, my guess is that it would depend on how well the machine conveyed "emotions." I place the word in quotes, because there is the question of whether a machine that is displaying emotions is also *feeling* them. But studies of people interacting with machines have shown that if the computer is given even a rudimentary face with eyes and eyebrows and a movable mouth, that people are readily engaged by a program that can maintain eye-contact and show basic emotional reactions to spoken words.

Should machines ever be developed that look very human-like, or perhaps were indistinguishable from humans unless they were opened and examined (such creations are called "androids"), and which had the ability to pass the *Turing Test* as well as the ability to display "human" emotions, it would be very hard not to take such a machine at its word if it told you that "he" or "she" was conscious and experiencing thoughts and self-awareness.

A number of science fiction writers have addressed the possibility of such machines one day walking and working among us—perhaps even becoming our friends or lovers. Such an occurrence would also raise many important issues. Should such machines be given their freedom? Should they be covered by the Bill of Rights? Should they walk among us as equals? The answers seem to hinge on whether or not they are "alive" in the sense that they are conscious entities. But there will never be a way to know other than to ask them, and that would not yield a certain proof. No, it will all have to be resolved in the political arena, simply because science can never hope to offer an answer. Can you imagine one or two hundred years from now, that the issue of slavery might once again be a dividing issue in the United States as well as the rest of the world? Robot rights could actually become a burning issue.

If it were possible to generate consciousness through the interaction of machine parts, would it then be possible to create a machine that could duplicate the combination of interactions that leads to the creation of your own personal consciousness? If so, could you cheat death by downloading yourself—your memories, attitudes, thoughts, into a machine, and thereby generate your consciousness anew—and in so doing, leave your old or dying body behind?

Computer scientist Ray Kurzweil addressed this issue in his book *The Age of Spiritual Machines.* Kurzweil asked what would happen if we could save someone from death by scanning that person with such detail that we could recreate him or her within a machine. If we could do so, the connections created within the machine would then be capable of generating a consciousness identical to that initially generated by the individual's own brain. Kurzweil then wondered concerning the machine's new "mind,"

> Is this the same consciousness as the person we just scanned?...The position that fundamentally we are our "pattern" (because our particles are always changing) would argue that this new person is the same because [his] patterns are essentially identical...the new person will certainly think that he was the original person...There will be no ambivalence in his mind whether or not he committed suicide when he agreed to be transferred into a new computing substrate leaving his old slow carbon-based neural-computing machinery behind. To the extent that he wonders at all whether or not he is really the same person that he thinks he is, he'll be glad that his old self took the plunge, because otherwise he wouldn't exist.
>
> Is he—the newly installed mind—conscious? He certainly will claim to be.

APES, DOGS, AND ANTS

All animals are equal but some animals are more equal than others.
—George Orwell, "Animal Farm"

Before returning to the great question, "What happens to us after we die," and how our conceptions of consciousness relate to it, it might be interesting to ask who else, other than human beings, might be conscious. I ask this question because if it were possible for our conscious minds to be recreated, we might wonder if such a recreation would also require a human brain. So we might ask, are apes or monkeys aware? Do dogs or cats think? Could an ant be conscious? Is there any way that we can know? And, as bizarre as it might sound, can we seriously ask if we might ever spend a lifetime as one of these creatures? (Frankly, our cats are pampered as though they reside in the Penthouse Suite at the Plaza Hotel, so one can only hope. As they say, "Dogs have owners; cats have staff.")

Of course, if we can't even know if our fellow human beings are conscious, there clearly is no way that we can hope to know if apes, dogs, or ants are. But we can look for hints. In fact, there is a set of fascinating little studies that might offer us a clue.

The first study of interest was conducted in the early 1970s. In this experiment, 5-month-old human infants had a small smudge of rouge placed on their noses, and were then set in front of a mirror in which they could see themselves. None of the infants seemed to recognize who the baby with the smudge on his or her nose was. As a group, they failed to recognize their own reflections.

When the same was done with 12-month-old babies, it was observed that they would often reach out and touch the mirror, and sometimes even look behind the mirror. But, like their younger counterparts, none seemed to know that the reflection was of him or her.

But, when the same procedure was tried on 20-month-old toddlers, the common response was for the toddler to look at the reflection, and then reach a hand up to his or her face and rub the nose. These older children were aware of themselves. They knew what the reflection was and they could now see and understand that there was a red smudge on *their* noses, of which they had been previously unaware.

In similar experiments, monkeys and baboons have had their foreheads unobtrusively marked with colored dye. The animals were then placed individually before a mirror. But even after thousands of hours of viewing, there were no actions observed that showed any self-awareness, at least none like those seen among the 20-month-old children.

However, when great apes were chosen for the experiment (in this case a chimpanzee and an orangutan, animals closer to humans genetically than monkeys or baboons), the results were interesting.* Both apes responded alike. After a short exposure to the mirror, they would look carefully, and then bring their hands up to their faces and rub the colored marks! They appeared to be consciously aware and to know who they were.

This experiment appears in some degree to test for self-awareness, something we often accept as fundamental to the basis of consciousness. Could these findings, then, mean that great apes, along with our own species, have the brainpower to generate a consciousness? Perhaps it does. Might these results also indicate that consciousness in humans is not

---

* Great apes, e.g., gorillas, bonobos, chimpanzees and orangutans, are distinct from monkeys. Great apes do not have tails; monkeys do. Great apes are also distinct from the lesser apes, e.g., gibbons and siamangs. Great apes differ less from humans, in a genetic sense, than do the lesser apes, indicating that the lesser apes broke off from the genetic line that eventually led to humans sooner than did the great apes.

something that is present at birth, but rather something that ignites a few months before our second birthday? Maybe. It might very well be that consciousness is something that appears once the brain reaches a certain stage of development, and not before. My guess is that a certain amount or brainpower is required before consciousness can appear. It might well be that only humans and great apes, once they get beyond infancy, possess a sufficient brain to generate a consciousness.

Experiments concerning consciousness in animals other than apes and humans have led to generally mixed results, although more encouraging results have been reported for porpoises and elephants. But even those that indicate some semblance of thinking don't show such thought to be particularly advanced or prevalent in most species. Also, it is very difficult to observe a thought. The only way is to do so indirectly. For instance, consider a typical study in which a dog is trained to push a lever for a doggie treat. It doesn't take too long for the dog to learn that! Now, I place a symbol by the lever, either a square or a circle. If the square is presented, the lever will pay off when pressed. But, if the circle is presented, pressing the lever gains nothing. As you might guess, after a time the dog will learn only to respond in the presence of the square since pressing the lever when the circle is present has become futile. At this point, psychologists say that the dog has learned to discriminate between the two stimuli, the square and the circle. In fact, experiments like this have allowed us to show that dogs are color blind, because if you use a blue and red circle (that have been balanced for brightness) as the discriminating stimuli, the dog will never figure it out.

When humans try the circle/square discrimination we see similar results (although it is best to use dimes or points toward a "good score" than doggie treats, but there are always the odd exceptions). People soon learn not to bother responding to the stimulus that doesn't pay off. What makes this experiment interesting is the switch-a-roo. In the switch-a-roo, the stimuli are suddenly switched so that responding in the presence of the circle will now pay off while responding in the presence of the square abruptly stops yielding a payoff.

When this happens with animals we see a gradual lessening of responding to the stimulus that now offers no payoff, and a gradual increase in responses to the stimuli that are now yielding results. These are known as learning curves. But humans show a different response, a sudden and abrupt change from not responding to the stimulus that didn't used to pay off, to fully responding to it. The differences are highlighted in Figure 7.

When we conduct interviews with the people who participated in the experiment, we discover that this abrupt shift in responding appears to be a function of an internal thought process, or insight. People will typically say that they suddenly realized that the contingencies had switched and that now the other stimulus was signaling a payoff. They'll report having said something to themselves such as, "Oh, I get it; they've switched it around the other way." Once they have come to this conclusion, they will alter their responding *immediately*. This sort of sudden shift is not seen in animals such as dogs or cats. Of course, we can't get inside an animal's head, but these sorts of experiments yield results that appear to indirectly enable us to observe thought in humans and its effect. Having insight typically involves the manipulation of internal symbols, thinking things over in one's mind—a sort of learning through imagined experience. Indirect observation of insight in the great apes has also been experimentally observed, and some of these studies go back quite a way. In fact, the first such observation was recorded by Wolfgang Kohler during World War I. At the time he was director of the Anthropoid Station of the Prussian Academy of Sciences on the Island of Tenerife off the western coast of Africa, where he spent much of his time investigating a colony of captive chimpanzees. He noted that one of his chimpanzees, named Sultan, had learned to reach outside his cage with a stick in order to retrieve a banana.

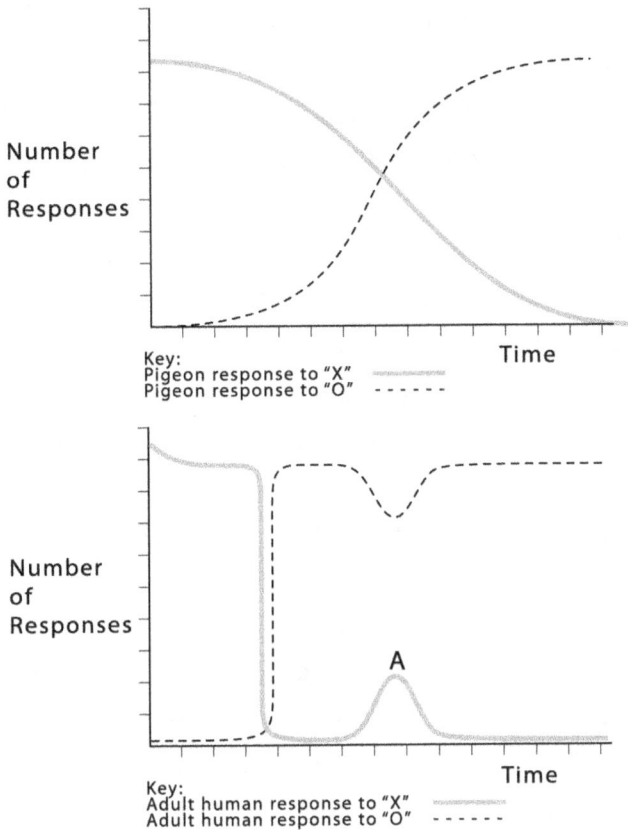

Figure 7. An idealized example of learning curves depicting a typical acquisition and extinction of a simple discrimination by both a pigeon and a human adult: Prior to these data being collected, both the pigeon and human adult had learned that if they responded to the "X" when it was present the response paid off, while responses to the "O" paid nothing. The data depicted in this figure were recorded after the contingencies were reversed so that the "O" would now pay off, while the "X" yielded nothing. Note that while the pigeon both accelerates its responses to the "X" and diminishes its responses to the "O" at a steady rate, the human's responses shift abruptly. The one brief spike in the human's responding [A] occurred when he briefly decided to check if the new shift was still in effect (e.g., whether "O" might have returned to paying off).

That particular action came as no surprise to Kohler. He knew that chimpanzees in the wild used sticks to fetch food, especially to gather termites, something they loved to eat. However, while pursuing his investigations, Kohler struck upon something more interesting. He developed a variation of the banana and stick problem. In this new situation, Kohler placed the banana out of the reach of a single stick and placed two sticks inside Sultan's cage. One stick was larger than the other and had a hollow opening into which the smaller stick could be inserted. Knowing this, you can most likely imagine the solution in your own mind; the smaller stick could be inserted into the larger one, so that together the two form one very long stick. With the long stick, Sultan would be able to reach the banana.

Watching Sultan in this new situation was very interesting. At first, the ape tried to reach the banana with the longer of the two sticks. When that failed, he pulled the stick back and tried again. Finally, he put the stick down and seemed to become moody and angry.

But then, after sitting quietly for some time while looking at the two sticks, Sultan suddenly appeared to realize something. His expression changed. He stood up and slowly approached the sticks. Then, he immediately picked up the larger one in his left hand and turned the hollow end toward him. He then picked up the smaller stick and inserted it into the larger one to make one long stick (see Figure 8). With the new, extended stick in his hand, Sultan went to the end of the cage and reached out toward the banana. He reached the banana with the stick and pulled it toward him!

Figure 8. Sultan, apparently displaying insight, as he assembles the two sticks.

It appears, then, that human beings and the great apes possess sufficient cognitive power to both recognize themselves in a mirror, and to generate internal symbols and manipulate them with enough agility to achieve insights. How well such ability correlates with the generation of consciousness is unknown. All I can do is make an assumption, and guess that great apes are conscious beings, but I can't be sure. As I noted earlier, I can't even be certain that other humans are conscious beings. But my guess is to include humans and great apes in the classification of beings who are conscious and who think. Ants, on the other hand, almost certainly don't think.

Ants, insects and other creatures with simpler central nervous systems, almost certainly don't have the cognitive power (either in quantity or quality) to generate consciousness. I can't put myself in an ant's head, of course, so I can't know for certain, but from all of the studies conducted with ants, there are none that indicate an underlying thought process. Ants do specialize in terms of the tasks they perform. They do appear to conduct organized war of a sort against invaders, and they do hunt, forage, and return valued items to their place of residence. Much of what they do looks purposeful. But Edward Wilson and many other researchers, who have spent lifetimes studying these amazing creatures, show us that these behaviors are driven for the most part by secreted chemicals. If you provide the right chemicals to the environments of ants, you can initiate various complex and specialized behaviors en masse. To the best of our ability to know these creatures, they appear to basically be "bio-robots." By this I do not mean to imply that they blindly follow programs and cannot learn, but rather that there might not be "anyone home" to actually think anything in the way that you or I might.

For example, if ants are placed on a metal plate that is cool, and then a heat element is turned on, the ants will scurry off the warming plate. But was that because they felt pain? I can program a computer with an attached armature that contains a temperature sensor to withdraw the armature, even to withdraw it with great force, should it encounter a temperature that might be harmful to it. But does that mean that the computer felt any pain? I doubt it (and I certainly hope not).

Without ever being able to actually know, I am going to assume that a certain quantity and quality of cognitive power, as reflected by the complexity of the neural circuitry of the creature being considered, is required for the generation of consciousness or even to be aware of pain in the way that you or I might. I believe that humans and the great apes are conscious. I think monkeys, dogs, and cats, pigs, horses, etc. probably aren't, or if they are, they have a rudimentary form of awareness. I believe that most creatures with less complex nervous systems—fish, birds, insects, lesser mammals, and so on, are not consciously aware. I might be wrong on all counts. I can offer nothing but my best guess.

The reason for considering this issue, as I mentioned earlier, is that if we are to consider what happens to us after we die, and if we are assuming that it is possible for one's consciousness to be recreated, we might wish to consider to just what sort of body your or my new consciousness might be relegated. It is not my intention to espouse any theory of reincarnation currently on the map, I only ask: If it is possible for your consciousness to reoccur, are there restrictions that would determine which sort of being you'd have to be in order to support it?

Could your consciousness be supported by the brain of an ape? (And I mean nothing personal by that remark). As it turns out, there is a range of possibilities, from narrow to broad. In the narrow view, your personal consciousness would be much attuned to your own brain, and slight alterations to your brain might make it impossible for *your personal consciousness* to exist within you. Furthermore, it would be unlikely that only but a few types of human brains, or perhaps only one type of brain, could ever support *your* consciousness. In the broad view, it might be that any healthy human brain or the brains of any other creatures that possessed sufficient cognitive processing power could support your personal consciousness. Also, this broader view might also allow that a machine with sufficient cognitive processing power might be able to support your personal consciousness.

We will never know the answer, of course, because we can never know who is, or who is not, conscious. Because of that fact, we will never be able to measure the parameters associated with the generation of consciousness. Perhaps someday, if an advanced computer suddenly claims to be conscious and aware, we can at least see at which point, and under which circumstances, it will make that claim. Then we might get a clue as to what sort of processing power, and what sort of arrangements of that power, will yield such a claim. Perhaps if we can then find similar arrangements in the human brain (and perhaps in the ape brain as well), we could then venture a better guess as to the requirements of a consciousness. But we will never know for certain.

It might also be possible someday to conduct similar research through genetic manipulation that would create new creatures that had successively greater cognitive abilities, and then see what they claim, although such an idea seems frightening to me. It is curious that such an idea would seem frightening to me while the idea of doing the same with a possibly sentient machine would not. Perhaps that is something I had better rethink. There is a long and ugly history of what people are capable of doing to those they consider sub-human, and after all of our experiences with truly mindless machines, it might be very easy for us to not have the appropriate respect for a machine consciousness should one appear, especially since it would have been a consciousness that we ourselves had built. If you don't agree with me, imagine the possibility that it is *your* consciousness that someday appears in the machine when a certain processing power is obtained. In that case, machine rights might seem a more pressing concern.

Of course, should a machine consciousness ever emerge, there would be many who would argue that it wasn't a real consciousness, but rather simply the machine just saying that it was conscious. That issue could never be resolved, but it might lend weight to those who wished to continue thinking of machines as sub-human.

Should such machines ever surpass us in intelligence, perhaps they will be able to educate us to better understand them in ways we have not considered. Or perhaps they will come to look down on us as "sub-machine." Numerous science fiction stories have dealt with such issues. But computer scientists like to respond, "Well, we will always have control because we will be the ones who can unplug them." Somehow, though, I can't take much solace in the fact that we would be counting on our ability to keep a superior being on a leash.

JUPITER AND BEYOND

When I heard the learn'd astronomer,
When the proofs, the figures, were ranged in columns before me,
When I was shown the charts and diagrams, to add, divide, and measure them,
When I sitting heard the astronomer where he lectured with much applause in
the lecture-room,
How soon unaccountable I became tired and sick,
Till rising and gliding out I wander'd off by myself,
In the mystical moist night-air, and from time to time,
Look'd up in perfect silence at the stars.
—Walt Whitman, "Leaves of Grass"

We have seen that if it is possible for one's consciousness to reoccur, and if we accept the broad view, then we may also argue that one's consciousness might reoccur in any creature with sufficient neurocognitive power to support it. In other words, if your consciousness did reoccur, it might not, of necessity, reoccur in the body of a *homo sapien*. Perhaps your consciousness could reoccur within the body of a great ape, or an advanced machine, or some other interesting creature.

In a finite universe, the broader possibilities, namely that your consciousness could be supported by a wider range of organisms or machine processors, increases the probability that your consciousness would reoccur simply because it would have more opportunities in which to do so.

If we are to make the assumption that your consciousness was generated by a particular brain organization or pattern, it makes logical sense then to assume that an identical, or perhaps similar organization, would regenerate this same consciousness, in much the same way that a certain element when excited by a certain energy, will always emit photons (particles of light) of the same wavelength. We then must ask the upsetting question, "Well, what are the odds of such a thing ever happening?"

As you recall, the last time we played the odds game and attempted to discover what might have been the chances of you or I having ever been born—we got nowhere. Some considerations lessened the odds, while others only lengthened them. This time, however, I am going to argue that it makes some sense to consider the odds of your consciousness reoccurring, because advances in physics, astronomy and cosmology indicate that the chances of such an occurrence seem to be trending all in the same direction, toward making it more likely, in fact, *much* more likely.

So far, we have considered a reoccurrence of consciousness as perhaps being possible within the confines of human brains or even those of the great apes, or perhaps within the circuits of advanced machines. But even if we take the most liberal assumption, and imagine that all of these sources are fertile ground for the reoccurrence of one's conscious mind, they still leave quite a limited number of possibilities. The number of people is limited to billions and apes far less. As for machines, who knows how many extremely advanced models could be built, but I am guessing less than the number of people.

But for us to assume that we might have a good chance of having our consciousness recreated, we might wish that there were many more possibilities for fertile ground in which such consciousness might take root; trillions upon trillions of possibilities, or perhaps far more. Actually, the more the merrier, especially if we consider the narrow view that consciousness might require a human brain in order to exist, or that your consciousness might even require a certain type of human brain.

But where would we find such resources, that is, neurologically powerful enough brains to support consciousness once we have exhausted human, animal, and machine possibilities? The answer lies beyond the Earth.

Now, at this point you might rightly fear that this book is about to take a bizarre turn and head off into outer space in search of aliens—and, to a certain extent, you'd be right. But it makes perfectly logical sense to do so. I won't be departing in the least from my efforts not to bring into play explanations for which there are no credible scientific backing. I am not about to point to any specific aliens, because I know of none. As far as I know, there are none. But if we look at the progress of modern physics, astronomy and cosmology (the study of the universe, not to be confused with cosmetology, the study of make-up and its application), you will see that the suggestion that we are the only intelligent species in the universe is most likely an absurd one. And, well, more brains equal more places for our consciousness to perhaps reoccur. In fact, if your consciousness once resided in a body that evolved on another world, and at that time you had read a book like this one, and that book referred to an alien into whose mind your consciousness might someday reoccur, it would be, unbeknownst to either you or the author of that book, referring to a strange-looking bipedal species on a distant blue planet that was called Earth by its inhabitants. To them, we are the aliens, whoever "they" might be.

At this point, I want to stress that I am not yet arguing that my view of a possible afterlife is, in fact, based on the concept of a repeating consciousness. It actually gets a good bit more fantastic than that. I only ask that you bear with me for a little while and hold onto that as a possibility. In a like manner, the foray that we are about to take into astronomy is also to lay the groundwork for a more stunning view that might not be apparent at the moment. So, for the next few sections let me take you from a time when people understood our universe to be composed of five planets, a sun, a moon, and a lot of stars all circling the Earth—to our current mind expanding and utterly dazzling comprehension of the modern universe(s). It is important for our journey to show you these amazing discoveries, so please indulge me as we continue along together. I trust that you will not be disappointed.

Perhaps the best way to begin our discussion of this "extra-terrestrial" topic is to state flatly that no life beyond the Earth has ever been proven to exist, although recently there are some hints of such a possibility. The idea of life on other worlds, however, is hardly a new one. The ancient Greeks believed in the possibility of a plurality of worlds inasmuch as atoms (a fundamental constituent of matter postulated to exist by the Greek Democritus) might be infinite in number and, therefore, could not have all been used just to create the Earth. Even though other worlds were not readily discernible to the ancient Greeks, they *reasoned* that such worlds might exist and have people living on them.

In Roman times, the poet Lucretius (99 B.C. – 55 B.C.) supported this idea and argued that other worlds must exist since the cosmos in all directions appeared to go on forever.

Aristotle (384 B.C. – 322 B.C.), the great Greek philosopher, took an opposing view. In his conception of the universe, Earth was in the very center of a series of concentric and nested spheres. He argued that there could not be other worlds as there was only one center, and that a second center would be disruptive, causing water and fire to be attracted to a second focal point. Aristotle's view was strongly supported by the well-respected Greek astronomer Ptolemy (approx. 85 A.D. – 165 A.D.). In the 4th Century A.D., St. Augustine concurred, arguing that if there were other worlds with other people they would also require a redeemer, and since the Bible describes only one Christ, there could, therefore, only be one Earth.

By the 12th century, following the Dark Ages, Aristotle and St. Augustine's statements had become well established as the final say in matters concerning the possibility of other worlds. It also seemed quite sensible to most scholars of the day, that if man were at the center of the universe, and if man were God's greatest achievement as many believed, there would not, *could not*, be other worlds. However, the questions created by the

Greek view of a plurality of worlds would not go away and continued to yield discussions that made for some discomfort among the clergy.

In response, over the course of the 13th and 14th centuries, the issue was generally resolved by conceding that if God had *wished* to do so, he most certainly could have created other worlds and numerous "centers" within a single universe, but that there was nothing in scripture to suggest that He had ever done such a thing. This apparently left Aristotle and St. Augustine in the clear. Human beings were at the center of the universe and *all things extraterrestrial* revolved about the earth imbedded in nested spheres of what was most likely the finest and clearest crystal imaginable (crystal appeared to be required, as it was the only substance of which anyone could conceive that might hold imbedded objects, such as the stars and the moon, securely so that they did not fall and yet be so clear as to be invisible to the eye). Questions were raised from time to time (such as what "shooting stars" might be), but not many, and when such questions were raised, they were generally ignored. Time passed slowly...then came the first of a series of great shocks.

In 1543, Nicolaus Copernicus wrote *De revolutionibus*, in which he showed that a heliocentric model that placed the sun at the center of a solar system, better predicted the motions of the planets than did the geocentric model espoused by Aristotle and Ptolemy. The Copernican view also gave rise to the conception of Earth as simply one of the six planets in orbit about the sun, which included Mercury, Venus, Mars, Jupiter, and Saturn. Quite naturally, if Earth was just another one of the planets, and Earth had people living on it; it was not unreasonable to wonder if people might not be living on the other worlds as well.

It was then that the brilliant Italian scientist Galileo Galilei turned his homemade telescope to the heavens and forever changed the world. In 1610, he announced his discovery that the planet Jupiter appeared to have four moons and that these moons, as he had observed over time, were in an orbit that *circled Jupiter*. This clearly did not fit with the concept of everything going around the Earth while imbedded in fixed nested spheres of the purest crystal. Perhaps the Earth wasn't the center of the universe, and perhaps man wasn't at the center of everything (oops, don't you just hate when that happens?) The Roman Catholic Church immediately forced Galileo to recant his observations with threats of violence and imprisonment. Galileo, an old, frail, and by then very frightened man, did so. The Joven moons, unswayed by church dogma, continued to silently orbit Jupiter. The church apologized for its error 369 years later in 1979.

In his writings, Galileo had also noticed that our own moon looked Earth-like, inasmuch as it appeared to have areas of land and sea (he mistook light and dark rock for land and water). Inspired by these observations, Johannes Kepler, a German mathematician whose calculations of planetary motion did much to support the heliocentric view proposed by Copernicus, hypothesized in 1610 that one crater of particular circularity might be an artificial construction produced by lunar inhabitants. In 1638, Bishop Wilkins wrote *Discovery of a World in the Moone*, in which he argued that there was life on the Moon and what the religious implications of such life might be.

In the light of speculation of life on the Moon and the other planets, it was a short leap to consider the existence of other solar systems that might also be harboring plants, animals, or even people. Among those speculating upon such possibilities were René Descartes and the well-respected Dutch astronomer Christiaan Huygens. However, without more powerful instruments, or a way to visit the other planets, the issue remained impenetrable. For the next few centuries, the idea of life on other worlds went dormant and remained unremarkable.

That all changed in 1895. Sir Percival Lowell, working with a powerful refracting telescope located in Flagstaff, Arizona, began spending great amounts of time looking at the planet Mars, especially during its times of closest approach to the Earth. Lowell was a good astronomer; there is no denying that. But if you have ever looked for any length of time at an object through a telescope, especially if that object is right at the limits of your resolution power, you might notice that you will just barely be able to see some things and not be quite able to see others, or perhaps you can see them for just a second, or perhaps not...your eyes water...you refocus the instrument...there was something...or was it the dancing of the Earth's atmosphere...no, it was there, perhaps if you looked a little harder...maybe if you made drawings and compared them over a number of days?

Something like this was happening to Lowell as he spent night after night examining Mars. He saw small dark patches, almost dots, scattered about the Martian surface. Over time, he connected the dots. He was certain that he could just make out ultra-thin lines connecting the dark patches; sometimes they were gone, but sometimes he could just barely see them. Being the well-trained scientist that he was, he trusted his own eyes.

What might the dark spots be? What could the thin lines connecting them possibly be? There are no natural geographic features that cause perfectly straight lines to cross a planet's surface. Only man-made structures do that. Could the dark spots be water? Might the thin lines be canals made by Martians to stave off an inevitable end that was to be the fate of their dying planet—a planet slowly losing its water?

Such speculation quickly fired up minds, no more so than when it reached the ear of the English writer H. G. Wells. Wells quickly imagined a dying Mars with advanced beings true to the Roman god Mars' namesake—a world of warriors. Wells knew that there were efforts underway on Earth to build aircraft, and many lighter-than-air ships had gone aloft. Might advanced machines someday traverse the distance from Earth to Mars? Or, more frighteningly, might they come from Mars to Earth? The opening passage from Wells' "The War of the Worlds" is written to terrify:

No one would have believed in the last years of the nineteenth century that this world was being watched keenly and closely by intelligences greater then man's and yet as mortal as his own; that as men busied themselves about their various concerns they were scrutinized and studied, perhaps almost as narrowly as a man with a microscope might scrutinize the transient creatures that swarm and multiply in a drop of water.

At the beginning of the 20th century, science was taking strides forward that were capturing the imagination, and Lowell's observations, reinforced by Wells' novel, fueled intense interest. For the first time, the consideration of life on other worlds became a common topic of discussion among the general public. Scientists speculated that it might be possible to reach Mars someday by using a rocket. The first hardbound science fiction series, the Marvel Books by Roy Rockwood (begun in 1901), included a story called *Through Space to Mars* in which Earthlings reach the Red Planet via a rocket ship (and once there, steal from the Martians and kill them all in jingoistic good fun because, after all, they were Martians and what else would you do but fight them?).

The idea of Martians became popular and was included in many stories and novels, including some written by Edgar Rice Burroughs, better known for his creation of Tarzan. Venus, too, became a probable place for intelligent life in the minds of many—sister planet to Earth in size, secluded in secrecy by a never-clearing cover of clouds. Perhaps the clouds kept the surface cool and protected the inhabitants from being nearer to the sun? Stories often depicted Venus as jungle-like, with swamps and steaming tropics, and often harboring masses of scantily clad women (after all, Venus is named for a goddess). Still it was Mars that held sway. Mars, red in color and named after a god of war just sounded scarier. Belief in life on Mars became commonplace in spite of the growing doubts most astronomers were beginning to express about the existence of Lowell's canals.

When Orson Welles made a radio broadcast of *The War of the Worlds* for The Mercury Theater in 1938, the nation was ready to believe such a thing was possible (perhaps in part because Hitler had everyone's nerves on edge over the prospect of another world war). Welles had the added misfortune of making his show sound realistic by interrupting dance music to bring his listeners a breaking "news" announcement. My mother's father had the extra misfortune of living in rural New Jersey, and so when Welles began his "news" of a Martian invasion by saying that he was broadcasting live from Three Corners, New Jersey (a tiny, obscure, but very real town in New Jersey that Orson Welles was nowhere near at the time), my grandfather knew exactly where it was. His response was to grab a shotgun and bar the door. Then he stood guard for a full hour, ready to return any death-ray zap with a blast of buckshot. The Martians never came and the family never let the poor man hear the end of it. From that day on, if anyone in the family disagreed with any of my grandfather's judgments or opinions, you could be sure that Martians were soon to enter the discussion. He took to drinking beer and mostly sitting by himself.

Even though *The War of the Worlds* broadcast turned out to be a Halloween radio program and not a real attack, people continued to search the skies for possible invaders from space, often turning unidentified flying objects into identified ones from other worlds. Following World War II, the United States Air Force began *Project Blue Book*, which was a serious investigation of UFOs, not because the Air Force thought that there were aliens about to invade, but because they wondered how many of these strange sightings might be aircraft of one sort or another originating in Russia. Many, however, took the official interest as a sign that there must be some truth to the idea that flying saucers were now among us, and interest and fear grew. Mars and Venus continued to be suspect.

Excitement about the possibility of life on Venus or Mars continued unabated until the mid-1960s and 70s, when those either hopeful or fearful of someday meeting a Venusian or a Martian received a cold dose of reality.

I can recall watching television in 1965 and waiting for the first close-up photographs to be transmitted back to Earth from the Mariner IV spacecraft as it neared Mars. I also remember hearing the scientists let out a collective groan of disappointment as the first pictures clearly showed the red planet to be a world of craters. Like the Moon, it had "dead" written all over it—no cities, no Martians, no death-rays, and not a canal in sight.

In 1975, the Soviet Venera 9 Lander descended through the thick clouds of Venus and came to rest on its surface. There it lasted for only a few minutes, quickly succumbing to a ground temperature of about 800 degrees Fahrenheit and an air pressure of over 30 atmospheres. Venus did not have swamps or jungles, or any women; it was more along the lines of Hell. The clouds hadn't kept the heat out; they'd held it all in. If anything was living there, it wasn't based on DNA.

The prospects of intelligent life in our solar system, aside from what dwelled on Earth, appeared to be at an end.

No one had any hope of finding life on Mercury, with its almost non-existent atmosphere and blazing heat, and the great gas giants beyond Mars offered few prospects. Jupiter, for instance, doesn't even appear to have a surface in the sense we might imagine one, but rather a hydrogen core crushed under such pressure as to perhaps cause the hydrogen to be in a solid metallic state. Had Jupiter been ten times larger than it is, it would have crushed the hydrogen to the point of fusion and ignited as a companion star to our sun. In fact, most stars we see at night are actually double stars. Ours isn't, but only because Jupiter was a bit too small to light itself.

Where, then, might we look for fertile ground for the generation of a conscious mind other then here on Earth if not among the planets within our solar system?

WORLDS WITHOUT END

Space is big. You just won't believe how vastly, hugely,
mind-bogglingly big it is. I mean, you may think it's a long way
down the road to the drug store, but that's just peanuts to space.
—Douglas Adams "The Hitchhiker's Guide to the Galaxy"

As you recall, both the ancient Greeks and Romans considered that the heavens might go on forever and that the number of atoms might be infinite. This was something that seemed reasonable; although there were scholars at the time who argued that while the numbers of atoms might be infinite, the number of stars might not be. This argument was solidified by the German astronomer Heinrich Wilhelm Olbers, who in 1823 reasoned that if the number of stars were infinite, the light from them all would be blinding. Because the number of stars was clearly finite, it was then considered not unreasonable to assume that perhaps the number of atoms was, too. But there was no way to know.

What can modern science tell us about the scope and size of the universe? Just how big is it actually, and what exactly is out there, and might any of what is out there be conducive to the formation of intelligent life?

For many centuries, the heavens were poorly understood and thought by most to be the abode of gods. No one knew what stars were, or what the "wandering stars" called planets might be. (In Greek, "planet" literally means "wanderer"). All that was known about the planets was that there were five of them, Mercury, Venus, Mars, Jupiter and Saturn. The view put forth by Copernicus was that the Earth should join the list as the sixth planet.

With the publication of Sir Isaac Newton's major work, *The Principia* (1687), it became clear that Copernicus had been correct and that we did indeed live in a heliocentric solar system that could now be understood in terms of gravitational attraction. Newton's equations for gravity explained and predicted the orbits of the planets about the sun.

For the first time there was speculation that the stars might be other suns. If so, it would mean that these "suns" were very far away because they were so dim when compared with our own sun. Thought also turned to the possibility that there might be other planets orbiting these distant suns, perhaps with living beings inhabiting them. It just seemed likely that there were more planets "out there." The problem was that there was no obvious way to explore such distant places.

Then things began to change. The first strong indication that there might be more "out there," came in 1781 when a German-born astronomer living in England named Wilhelm Herschel announced the discovery of a new planet in our own solar system! Using a homemade 6¼-inch telescope, Herschel began examining the night sky. There, among the stars of the constellation Taurus, he found a slowly moving object that he at first mistook for a comet. But the comet had no tail. With a change of lenses, Herschel raised the power of his telescope from a magnification of 460X to 932X. At the higher power he now observed a pale green disk—a new planet!

When King George was informed of this stunning discovery he immediately made Herschel the King's Astronomer and added a hefty 200 pounds per annum to go with the title. In gratitude, Herschel (now going by the name William, not Wilhelm, to emphasize that he was an Englishman) named the new planet *Georgium Sidus* (George's star), and until 1850 it was listed in catalogues as the planet "Georgian." European astronomers, however, preferred the name Uranus, after the god who was the father of

Saturn and the supposed discoverer of astronomy. Slowly, the name Uranus supplanted Georgian.[*]

The news of the discovery stunned the world. It clearly meant that there really were more worlds "out there" beyond what was known. How many there might be, either in our own solar system or perhaps in others, was anybody's guess.

As you might imagine, Herschel immediately became world famous. Notables from all about descended upon his residence in hopes of seeing the new planet for themselves. Lords and ladies, often unschooled in even the barest essentials of astronomy, would sometimes request to view the planet during the day, or at a time when it had not yet risen, or after it had set, or on English nights when the cloud cover was as thick as pea soup. Frustrated by his frequent uninvited guests' lack of understanding concerning the requirements for viewing the new planet, and their expressed disappointment when the prerequisites were explained to them, Herschel had a lantern hung some distance away high in a tree that could be lighted on evenings when viewing the real planet was otherwise impossible. The green lantern, when observed slightly out of focus, yielded all the hoped for "mmmm"s and "aahhhhh"s one might desire from his unsuspecting visitors. One can only imagine how much Herschel must have been enjoying himself on those nights.

In 1792, Jean-Baptiste-Joseph Delambre, a French mathematician, computed tables for the positions of the planets including the new Uranus, basing his work upon Newton's laws of gravitational attraction. He then published his findings in his opus, *Tables du Soleil, de Jupiter, de Saturne,*

---

[*] Interestingly, on a very clear and dark night, Uranus is visible to the naked eye. But the ancients never took notice of it, or if any did, they did not notice that over a sufficient period of time it, too, could be observed to "wander."

d'Uranus et des satellites de Jupiter. However, there appeared to be strange discrepancies between Delambre's predictions for the orbit of Uranus and its actual motion through the heavens.

The director of the Paris observatory, Alexis Bouvard, struggled to make sense of these discrepancies. He recalculated the effects of the gravitational pull exerted by the other planets, but failed to find an answer that could explain the differences between what was predicted concerning the orbit of Uranus and what was being observed. In 1821 he speculated that the answer might lie in "some foreign and unperceived cause, which may have been acting upon [Uranus]." Perhaps, it was thought, there might be yet another planet out there. Others argued that perhaps Newton's inverse square rule for gravitational attraction broke down at long distances.

The rigor of Newton's laws notwithstanding, the race was on to find the new planet, if it existed. John Herschel, son of William Herschel, referred to the new world suggested by the data at a meeting of astronomers in England. He said, "We see it as Columbus saw America from the shores of Spain. Its movements have been felt, trembling along the far-reaching line of our analysis..."

Astronomers and mathematicians in France and Germany joined in the search as well. The race was a very close one. One English astronomer, William Lassell, lost the chance to be the discoverer because he sprained his ankle. During the 24 hours he remained in bed and was unable to attend to his telescope, his maid misplaced the paper upon which was written the mathematically derived coordinates showing where in the sky the new world might be found. Without the paper, Lassell had no idea where to point his telescope. By the time he finally reacquired the calculations it was too late. In the interim, the German astronomer Johann Gottfried Galle, relying on data provided for him by the French astronomer and mathematician Urbain-Jean-Joseph Le Verrier, made the discovery of the new planet on the night of September 23, 1846. Neptune, as the new world came to be called, appeared to be a sister planet to Uranus, another pale green gas giant almost identical in size, but nearly twice as far from the sun.[*]

In many ways, the discovery of Neptune in 1846 was a far more important advance than the discovery of Uranus. Rather than accidentally discovering a new world, as was the case for Uranus, a new world had been deduced to exist based on scientific calculation, and *then* found. Theory had preceded the discovery of a new object, something we expect and take for granted now, but wholly novel at the time.

Following along the same line of reasoning that led to the discovery of Neptune, Percival Lowell took a short break from his studies of Martian canals in 1905 to carefully calculate the orbits of Uranus and Neptune, and came to the conclusion that there were still some unresolved gravitational effects to be explained. Lowell then predicted the existence of a 9th planet in our solar system that he called "planet X."

Excited by the prospect of discovering a new planet, Lowell spent two years searching an area of the night sky that he thought most likely to yield results, but found nothing. Seven years later in 1914, Lowell redid his calculations and spent another two years photographing a very small area of the starry sky, once again in search of the elusive planet X. He found nothing. He later acknowledged that not finding "X" was the sharpest disappointment of his life.

---

[*] Amazingly, Johann Gottfried Galle was not the first person to actually see Neptune with his own eyes. That honor belongs to Galileo Galilei on the night of December 28, 1612! Galileo was observing Jupiter at the time and marked Neptune down as a dim 8th magnitude background star also in his field of view. His telescope did not have the power to resolve Neptune as a disk, but it was strong enough to make the distant planet visible.

Astronomers have gone back and reexamined Lowell's photographic plates from the 1914-1916 surveys and have discovered an ultra-faint dot that had gone unobserved by Lowell. Ironically, the man who saw canals on Mars when there were none, failed to see Pluto when it was there.

In 1929, Clyde W. Tombaugh joined the team of astronomers at Lowell's Flagstaff observatory. His job was to search for planet X, and to do so he brought to bear a new device, called a blink microscope. The idea behind a blink microscope is quite simple. Two identical areas of the night sky are photographed some time apart. They are then placed in the blink microscope side by side so that an observer looking into the instrument will see the two images overlap perfectly. However, one image will be seen with his left eye, and the other with his right. The images are then alternately blocked at a rapid rate, causing a "flicker" so that the observer sees first one image and then the other. The value is simple. Stars in both photographs will be at identical locations since we are talking about two images of the same place in the sky. Even while the instrument alternates between images, the stars will appear not to move. But, if a moving object were in one photo, but not the other, it will appear to blink, (or, if still in the second picture but in a different spot, "jump" from one place to another). The tiny ultra-faint dot that had barely changed locations in Lowell's photos, and which was easily mistaken for a steady and very faint background star, would have been observed to jump had Lowell been looking at successive photos of that area through a blink microscope.

With this device at his disposal, Tombaugh wrote, "... on the afternoon of February 18, 1930, I suddenly came upon the images of Pluto! The experience was an intense thrill, because the nature of the object was apparent at first sight." Once again, it was shown that there was more "out there," beyond what had been known.

We have since expanded our knowledge of Pluto, a distant and cold little sphere named for the god of the underworld. From the start it was an oddball. Its orbit about the sun is inclined 17 degrees from the orbital plane of the other planets. Pluto's orbit is also more elliptical than the orbits of the other planets, even bringing it at times within the orbit of Neptune. In fact, from 1979-1999, Pluto was closer to the sun than was Neptune.

Pluto is now understood as perhaps one of the larger objects in the Kuiper Belt, a disk-shaped region of millions of small icy bodies that circle the sun between the orbit of Neptune and extending out to about three times the distance of Pluto. The Kuiper Belt is believed to be the source of short-period comets (comets that orbit the sun on a fairly regular basis, such as Halley's Comet). Other large objects within the Kuiper Belt have been discovered besides Pluto. Among these is Pluto's moon, named Charon, which is about half the diameter of Pluto, Quaoar, a distinct and separate icy body also about half the size of Pluto, and Varuna, another separate body about 40% the diameter of Pluto. Even further away (about three times further from Earth than Pluto), is the planetoid Sedna (named for the Inuit goddess of the ocean). It is about ¾ the size of Pluto, and has a surface temperature of about 400 degrees below zero Fahrenheit. Sedna takes 10,500 years to make a single orbit of the sun.

These discoveries have led many astronomers to conclude that Pluto should not be considered a planet in its own right, but rather be thought of as the largest comet nucleus yet discovered. Still, there is sentiment to call it a planet for historical reasons, but most astronomers believe that it is just a matter of time before objects in the Kuiper Belt larger and more distant than Pluto are found, which will pretty much put an end to Pluto's status as planet. The point is Pluto is not out there all by itself; it has a lot of company.

Gravitational studies have also indicated that there might be more to our solar system beyond the Kuiper Belt. A great sphere of dark icy chunks might surround the sun at about ¾ of a light year distant, roughly

1300 times further away from the sun than Pluto. This great surrounding sphere of debris is known as the Oort Cloud, named for Jan Oort who noticed that the orbits of long-period comets indicated an origin of about ¾ of a light year distant. The Oort Cloud might actually reach in toward the sun further than expected and some have argued that Sedna might actually be better placed as an object at the very beginning of the Oort cloud, rather than as a distant object in the Kuiper Belt.

Clearly, the solar system is filled with more "stuff" than anyone had previously imagined, although as far as our solar system is concerned, intelligent life appears to exist only on the Earth. (And speaking of intelligent life, I have to admit to having a hardy laugh when I discovered that some astrologers have rushed to include Quaoar, Varuna, and Sedna in their astrological charts, arguing that the omission of these objects might be the reason that horoscopes haven't always been exactly accurate in the past. So, by all means, if you have your horoscope done, be sure to visit an astrologer who includes these trans-plutonian planetoids or else, who knows, you might not get an accurate result!)

If we wish to look for intelligent life that could support a consciousness, (even one that would rely on a horoscope—I'm still laughing), it appears that we will have to search beyond our own solar system. Is there intelligent life somewhere out there? Of course, we might first ask if there is *any* life out there. It might be that not only intelligent life, but all life, is confined to Earth. With this in mind, scientists are interested in looking for any signs of life beyond the Earth. There have been indications.

The first tantalizing indication of life on another world was discovered in 1996, inside a rock known as ALH84001. While the name of this rather small unassuming rock might at first appear to look like the number on a license plate, it actually stands for rock number one (001), found in the Allen Highlands of Antarctica (ALH) in 1984 (84). The rock is an orthopyroxene cumulate, containing minor amounts of plagioclase glass, chromite and carbonate. While none of that will make much sense to anyone but a geologist, suffice it to say that rock ALH84001 is among the few rocks on Earth ever found that certainly had its origins on Mars. The rock is a meteorite, formed an amazing 4.5 billion years ago (by evidence of its crystallization time), making it one of the oldest rocks ever found. We know that it traveled through space for 16 million years because of the evidence of cosmic ray exposure that accumulated on and within the rock. It also contained oxidized iron within the chromite of the rock, which is specific to Mars. The history of ALH84001 is written clearly within its confines. It was formed on Mars 4.5 billion years ago, ejected (probably by either volcanic activity or meteor impact on Mars), sailed through space for 16 million years, and landed on Earth in Antarctica in about 11,000 B.C. (based on the evidence of time it was exposed to Earth's atmosphere).

What made ALH84001 truly stunning are a series of very tiny (about 50 nanometers in length) tubules found within the rock. One of the scientists working on ALH84001 took photos of the strange tubules. He was trying to come up with a geological explanation for why such structures would be within the rock and wanted to examine them in detail. While at home, he idly showed the photos to his wife, a biologist, who asked him what they were, since, as she said, she was unfamiliar with those particular fossilized *bacteria*! Talk about getting a jolt!

The possibility of fossils from Mars sent a quick shockwave around the world and even drew a press conference from then President Clinton about the high probability that life had once existed on Mars, and might still exist.

If the tubules are fossilized bacteria, they are probably too small to be from Earth, as no known bacteria on this planet are that tiny. Biologists have agreed, however, that there is no obvious fundamental reason why bacteria couldn't evolve to be that size. Still, there are geologists who think that there might be ways in which natural geological activity could yield forms and shapes that happen to look a lot like fossilized bacteria. More samples from Mars will be required before we can know, and several Mars lander missions have the investigation of possible life, current or fossilized, as their goal. The possibility of life on Mars has been further stimulated by the discovery of a frozen lake of water ice at the Martian equator the size of the North Sea, and by indications of current flowing water as observed by the Mars Orbiter and landers.

But when it comes to the search for life in our solar system, Mars isn't the only place of interest. Years ago, when Galileo turned his telescope toward Jupiter and observed the giant planet's four largest moons, he noted that they all looked alike—each one appeared as a little point of light. Today we know that these four great "Galilean" satellites—Io, Ganymede, Callisto, and Europa—are remarkably different from one another. Io is volcanic and so amazingly active that during a Voyager flyby mission photographs actually captured a great volcanic eruption in progress! Ganymede, the largest moon in our solar system, and large enough to be a planet in its own right had it been in orbit around the Sun rather than Jupiter, looks similar in many ways to our own Moon. Callisto, the furthest large moon from Jupiter, has the highest density of impact craters of any moon in our solar system, but possesses no volcanoes or even any large mountains. And then there is Europa—with its bright surface and young icy crust, a crust cracked in thousands of locations indicating that just beneath that fragile covering might lay an ocean of liquid water. Warmed by the constant tugs and strains placed upon it by the powerful Jovan gravitational field, Europa might have a great ocean, and perhaps life. There are plans underway to send a lander capable of drilling down through Europa's icy surface to explore such a possibility.

Finding life anywhere else in our solar system would be a stunning discovery. It would clearly mean that life was not a freakish occurrence, and that it is probably scattered throughout the universe. If there is life out there, especially if it is in abundance, we might also reasonably expect some of it to be intelligent.

It is too bad that there aren't other worlds in our solar system like Earth. But that doesn't mean that there aren't other Earth-like planets out there—somewhere. The search for planets outside of our solar system has become an exciting area of research., As of January, 2008, numerous planetary systems other than our own solar system had been discovered, containing at least 271 planets. As a general rule, these planets tend to be Jupiter-sized, or larger. That makes sense, since the largest planets are the easiest to spot first. Earth-sized planets will take a greater effort to find and will most likely require better instruments for detection. But the very fact that so many other planets have been discovered orbiting nearby stars, shows us that our solar system is not unique. Planets orbiting stars is, perhaps, the rule rather than the exception. In any case, it certainly isn't rare.

Perhaps, someday, a large orbiting interferometer will detect an Earth-like planet in orbit around a star 70 light-years away, or so. Perhaps our children or grandchildren will look upon a photo on their schoolroom walls showing a distant blue planet with scattered white clouds—a planet in orbit around another star—displayed alongside maps depicting the voyages of Columbus.

If you go out at night and look up into a starry sky, and think that many, or possibly most of the stars you see, have planets in orbit about them, you have to wonder if someone else isn't out there—perhaps looking back at you and wondering the same thing.

It is really remarkable to go out on a clear dark night away from city lights and simply look up. The number of stars is breathtaking. If it is a really dark and clear night, you should also be able to see the Milky Way, a wispy uneven band of hazy light that stretches across the sky.

The Milky Way is the galaxy in which we live. Actually, *every* star that you can see with the unaided eye in the night sky is in the Milky Way

Galaxy. The wispy band of cloudy light that we call "The Milky Way" is simply our spiral-shaped galaxy viewed edge-on (see Figure 9). Use a good pair of field glasses (they will work better than a telescope for this demonstration) and take a careful look at that hazy band of light. What you will see are stars, millions of them. It is one of the most beautiful and amazing sights I have ever seen and well worth your time. You don't even have to be in an extremely dark place. If you can see the Milky Way, that's good enough—just grab a good pair of binoculars and have a look. It's the best free show in town and, for some, a nearly spiritual experience. The hazy light of the Milky Way itself comes from all of the stars that are too dim to make out individually with the unaided eye. It is also interesting to consider that the center of the Milky Way would been seen in the night sky as a deep red glow as bright as the Moon if it weren't for the fact that so much interstellar dust near the galactic center was blocking our view of it.

The Milky Way is about 100,000 light years in diameter, meaning that it would take 100,000 years to cross it at the speed of light. It contains an astonishing 200 billion stars (give or take a few). That's 200,000,000,000 stars! Perhaps half of these stars have planets in orbit around them. How many "Earths" might be out there is anyone's guess. But since there is one here, I am guessing that it is not all that rare, although I could be very wrong about that. Still, 200 billion stars is a big number and might well imply the existence of a trillion planets. I would guess that's enough planets among which to find a few "Earths."

Are we really the only intelligent life in the galaxy? Are we the only creature across a trillion planets able to generate a consciousness? Maybe. As Enrico Fermi, the leader of the team that built the first atomic pile, used to ponder: "If technological civilizations are commonplace and moderately long-lived, the galaxy should be fully colonized and we should certainly have heard from them. But there is only a great silence—why?" This query has become known as *Fermi's Paradox.*

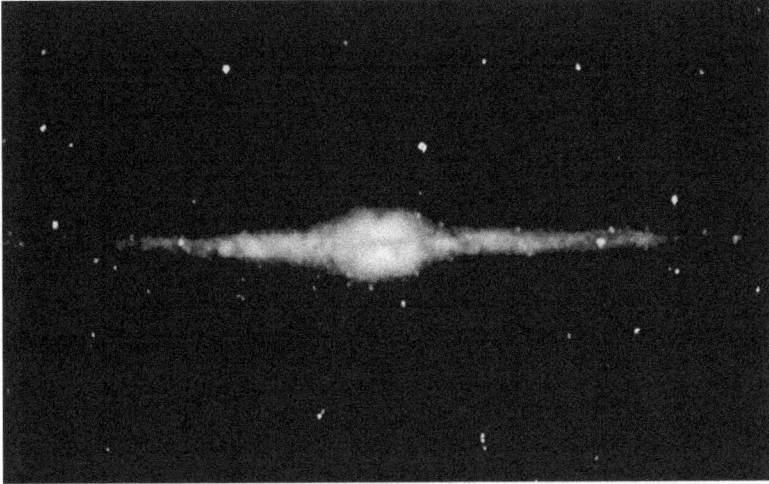

Figure 9. This image of our galaxy, the Milky Way, was taken with NASA's Cosmic Background Explorer (COBE)'s Diffuse Infrared Background Experiment (DIRBE), one of three COBE scientific instruments. This never-before-seen view is a combination of data gathered with DIRBE at intervals within the first six months in orbit and released in April 1990. It shows the Milky Way from an edge-on perspective with the galactic north pole at the top, south pole at the bottom and galactic center at the center. The picture combines images obtained at several near-infrared wavelengths. The dominant source of light at these wavelengths is stars within our galaxy. Even though our solar system is part of the Milky Way, the view looks distant because most of the light comes from the population of stars that are closer to the galactic center than our own Sun.

Credit: NASA, COsmic Background Explorer (COBE) Project

Fermi asked a good question. So, where in fact are these intelligent extraterrestrials and why haven't we heard from them? There are a number of possible answers, some of them quite intriguing.

One that is often heard is that we *have* had contact with them and that they have already visited us. These claims tend to center about various UFO sightings, alien abductions, and Area 51. I personally don't believe any of it and do not think aliens have either visited or contacted us. I find it suspect that these "aliens" never seem to land on the White House lawn, but rather always choose obscure landing sites inhabited by residents who appear to enjoy a good probing. Once again I recall Carl Sagan's admonition that extraordinary claims require extraordinary evidence.

As far as a government conspiracy to keep the truth about alien visitations from us goes, well, anyone who works for the government can tell you that if more than about six people know a secret, it won't stay one for long. You might say, well, what about the *Manhattan Project*? And I would answer that that makes my case, as Stalin often found out about the ongoing atomic discoveries before Oppenheimer did. Then there is an opposing view that is somewhat depressing. This view is that in this great and huge universe we are totally alone.

But these two are not the only possibilities. There is also a cynical explanation for the *Fermi Paradox* that goes as follows: Civilizations develop over long periods of time during which they don't have the technology to transmit electromagnetic signals. Finally, they discover radio and the transmissions begin, weak at first, but growing stronger as their technology improves. Very shortly thereafter (relatively speaking) they discover what one can do with uranium. Not very long after that, the radio and TV signals stop—. We haven't heard from them because there is a very tiny window allowing transmission from any given civilization before it blows itself to bits, and we'd have to be lucky in our timing to catch it. It's a horrible thought, and I can only hope not a correct one.

I personally prefer the technical explanation. Those espousing this view argue that there are many signals out there, but that our equipment or technique is currently inadequate for the job of detection. When you consider just how many stars there are, and then examine how many of them have been scanned by radio-telescopes (large dish antennae that can receive distant signals in the radio bandwidth), you will be dismayed to discover that it is an infinitesimal fraction of the number of stars actually out there, and that the sensitivity of the instruments that we are using is often sorely wanting. Not only that, but we can only guess as to which wavelengths other beings might use to broadcast so it is hard to know what to look for when we actually do look. Perhaps they aren't even likely to use radio, but rather rely on laser light, something we have only looked for very recently.

A good analogy can be seen outside my office window. I have a nice view of the woods on the other side of a country lane. Science tells us that there should be insects in the woods—but no matter how long I look at the woods through my window, I do not see any. I mean, if they are supposed to be there, then where are they? It's a paradox!

However, anyone walking through the woods who heard that I doubted that there were insects there because I never saw any from my window would have a good laugh. The fact is that the woods are teeming with all sorts of crawly and airborne critters—and there is no paradox. If I took a powerful enough instrument (perhaps a good pair of field binoculars) I might spy an ant crawling up a tree trunk. This could be the answer to Fermi's question—crappy detection.

Another possible answer, and a very fascinating one, is often called the Deification or Transcendental Solution. The argument is as follows: Civilizations gain technology and begin to explore the heavens. They imagine that space is the final frontier and they can now dare to go where no one has gone before. But soon they discover that not only can't the speed of light be surpassed, it can't even be approached—which is a shame since everything "out there" is so very, very, far away. Soon thereafter they discover that outer space is not the final frontier, but that inner space *is*. In other words, they discover that virtual reality, which started out as a video game amusement, is in fact becoming, with each additional technological advance, very, very, cool. Soon they are playing experiences directly into their heads, experiences that cannot be discriminated from actual experiences. As you might imagine, this all becomes quite popular. Within a few hundred years, these civilizations have altered their worlds so that robots do all the maintenance, and the people are free to hobbit themselves away in body preserving cocoons (ala *The Matrix*) while happier times are played out in their heads. And, should any become bored with what they are doing, they just erase their memories so that what's old is new again.

Perhaps some of the more discriminating among them have even found the one best experience they can ever have and are looping it over and over with a memory erasure following each experience so that the experience is always new. Such a loop might rightly be called a "heaven loop" since it is as close to the idea of heaven as any planet-bound individual is likely to get. If you don't agree, just consider the common descriptions of a "heaven" that have been the product of many cultures. They all have aspects in common. They typically describe a place where all beings live in untarnished joy, maintaining a state of perfect happiness forever and ever. Perhaps it sounds horrifying, but I don't know; think about it. Whatever you do throughout your day, no matter what it is, it's not likely to be as good as a heaven loop.

Anyway, if this is what is going on, and the fate of most or perhaps all intelligent beings is to eventually send themselves to "heaven" as demi-gods, and basically party 'til they drop; they would, in the words of Garbo, "...vant to be alone." These folks won't be broadcasting anything. In fact, they will most likely place their planet into some sort of stealth mode so that no one is ever likely to find them and crash their party.

If all of this sounds fanciful in the extreme, I only ask that you stop and think about it. If we humans don't have some great war and blow ourselves to kingdom come, what would ever stop our inevitably (sliding, striding) down that path? How many wonderful experiences can you dream of having that are either impossible, or simply never going to happen to you? What would you be willing to do to obtain those experiences if they were suddenly offered as a genuine option? Nope, the Transcendental Solution is a definite possibility!

But still, my guess for what that is worth is that we just haven't been looking very hard. The search for extraterrestrial intelligence, or SETI as it has come to be known, is often confused in the minds of lawmakers with UFOs, wacky folks, and kooks. To many members of Congress, voting for SETI has the ring of something that might be used against them in a campaign, so funding has not been readily forthcoming for what might well be one of the greatest adventures possible. All is not lost, however. Paul Allen (co-founder of Microsoft) and others, with the support of The University of California at Berkeley, have funded a $25 million radio array in California that is fully dedicated to SETI. Some people might scoff, but should we ever make verifiable contact with another civilization, it would without doubt be the greatest discovery of all time.

WHERE DOES IT END?

Many worlds, so much to do,
So little done, such things to be.

—Alfred, Lord Tennyson

In 905, the Persian astronomer 'Abd Al-Rahman Al Sufi described a fuzzy patch of light in the night sky. It is still visible today with the unaided eye in the constellation of Andromeda. There is little doubt that it had been observed by others long before 'Abd Al-Rahman Al Sufi laid eyes upon it, but he is credited as the first to make a written note of it. In 1612, the German astronomer Simon Marius became the first to report a telescopic sighting of the fuzzy patch. Through his small, low-powered telescope it appeared to him to be a slightly larger version of the same fuzzy patch of light than he was able to discern with his naked eye. Larger instruments would be needed if one wished to see more.

In 1923, Edwin Hubble (for whom the Hubble Space Telescope is named) turned the largest telescope of the day, the 100-inch Hooker Telescope at Mt. Wilson Observatory in California, toward this odd fuzzy patch. At the time, astronomers knew that the patch of light was not a star of some sort, because spectral analysis of its light did not match that of a star. Oddly, the light from the object seemed to give results that were inconclusive because the spectra from it ranged all over the place. It appeared that just about every sort of light was coming from it. Larger telescopes resolved the issue by showing that the fuzzy patch (called a nebula at the time) was really a grouping of a great number of stars in an amazing spiral shape. This explained the varied light spectra—light from the object was an amalgam of the light from a huge number of stars. Whatever sort of star collection it was, it certainly was the most unique feature in the Milky Way Galaxy.

At the time, Hubble was interested in a certain type of star known as a Cepheid variable. These stars were of special interest to him because of a peculiar attribute that they all shared. Cepheid variables have used up their supply of hydrogen fuel and have begun to fuse helium and other light elements. They also tend to be of a certain size. This combination of size and helium fusion causes an interesting thing to happen. When the star's fuel burns, the pressure of the resulting thermonuclear reaction causes an expansion, and a consequential cooling of the star. Shortly following this expansion, the star's gravity causes a contraction that results in a greater burning of fuel, causing the star to heat greatly and expand once again, which leads to a repetition of the expansion/contraction process. It is almost as though the star is breathing with regular inhalations and exhalations. To astronomers, the process shows itself as a brightening and dimming of the star at regular intervals (usually lasting between 1 and 50 days depending on the star, with each star maintaining its own regular interval rate).

The first of these Cepheid variable stars to be observed was delta Cephei, a star close enough to our own solar system to show parallax motion as our Earth orbited the sun. Parallax motion is easy to understand. Simply look at a close object (such as your index finger held before your face) first with one eye and then the other, and you will see that your finger appears to change position relative to objects behind it as you change from eye to eye—it seems to jump back and forth. The same is true of nearby stars. Looking at the star in spring and then again in fall (imitating a glimpse through different eyes set some distance apart), *nearby* stars will appear to move relative to the background of stars that are more distant. This is parallax motion. Using this technique, we can tell how close nearby stars are. Unfortunately, this method is of little use for stars that are more distant.

Because Cepheid variables show a very consistent brightness range, once we know how far away some of them actually are (through the parallax motion of the nearby ones) an interstellar "ruler" can be created. We can then estimate the distance to Cepheid stars beyond parallax range, simply by measuring how bright and dim they appear when they oscillate.

When Edwin Hubble observed the strange fuzzy patch of light in the constellation *Andromeda,* which he knew to be a spiral-shaped collection of stars, he was excited to discover that the Hooker instrument was powerful enough to resolve some of the stars in the collection as Cepheid variables. He quickly began to calculate the overall brightness of these variable stars based on the power of the 100-inch telescope. The result must have sent chills down his back. The entire Milky Way Galaxy is 100,000 light years across, but these Cepheid variables were 2.36 *million* light years away! This wasn't a strange spiral collection of stars in our galaxy—it was *another galaxy*! Up to that point it had been assumed that The Milky Way was it—one galaxy equals one universe. But now here was another entire galaxy, separated from us by a dark, void, intergalactic space, a galaxy as it turned out, 50% larger than our own, a galaxy with perhaps 300 billion stars in it! It was soon named The Andromeda Galaxy, and not too long after that, it was discovered that it wasn't the only other galaxy out there. Hubble quickly demonstrated that other galaxies nearby, now known as The Local Group, were also independent star groupings. In short order, finding galaxies became the rule, rather than the exception.

How many galaxies are there in the universe? No one is absolutely certain, but the total number appears to be astonishing. In 1996, the Hubble Telescope, in orbit above the Earth, was used to help give scientists an idea as to the total number of galaxies that might be in the universe. The telescope was pointed at an area of space that was considered fairly typical, and then the power of the scope was cranked all the way up. In effect, the Hubble was now focused on a piece of sky that could be hidden behind a dime at a distance of 75 feet. The photograph in Figure 10 shows what was found there. In the photo there are a few individual stars which are in our own galaxy and, in terms of galactic distances, quite nearby. But far more amazing is the bewildering collection of over 1,500 *galaxies* that appear in the picture.

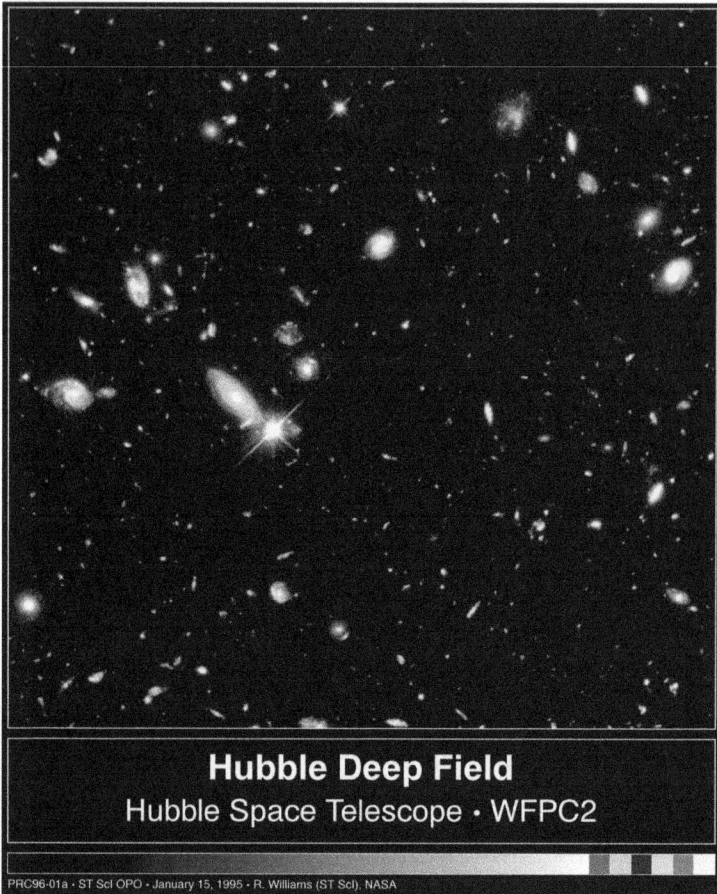

Figure 10. This deep space image captured by the Hubble Space Telescope encompasses a portion of the night sky small enough to be hidden behind a dime at a distance of 75 feet. In the photo there are a few individual stars which are in our own galaxy, and over 1,500 galaxies. This discovery, and other deep space images, has led to the rough calculation that there are at least 125 billion galaxies in the observable universe.

Based on such observations the current guess as to the number of galaxies in the universe is somewhat in excess of 125 *billion*! That probably puts the total number of stars in the known universe at perhaps 10 septillion (give or take a few). That's 10,000,000,000,000,000,000,000,000 stars! That number is so huge that it is difficult to make sense of it. Perhaps I can put it another way. Ten septillion stars mean that there is one star in our universe for every grain of sand to be found on all the beaches and deserts of the Earth, including the mighty Sahara. With that many stars, one can only guess at the number of planets, even Earth-like planets that might be out there.

The more Earth-like planets that there are, the greater are the chances that there will be other beings in the universe capable of harboring a conscious mind. In keeping with the argument that your conscious mind is the product of a certain brain combination, or combination of *something* (or else there would be no difference between you and other people), we might also assume that the chances of your consciousness reoccurring will increase dramatically with the addition of other worlds filled with beings into which your consciousness might reappear. With perhaps 50 septillion planets from which to choose, I would imagine that there can be no way we are alone in the universe, or that our planet is the only one supporting sentient beings.

Maintaining this line of thought, the more planets capable of sustaining life, the better. Amazingly, the planetary systems associated with these septillions of stars might not be the only places where intelligent life might reside. There might yet be more stars and planets out there, *many,* many more.

In recent years, astronomers and cosmologists have come to refer to the universe that we can see and observe as the *known universe*—as in, "There are about 10 septillion stars in the known universe." The question is, which parts are unknown, and how many stars and planets might be hiding there?

"MY GOD, IT'S FULL OF STARS"

But the star-glistered salver of infinity,
The circle, blind crucible of endless space,
Is sluiced by motion,—subjugated never.

— *Hart Crane, "The Bridge"*

Our Universe is expanding. The analogy often presented is that of dots (representing stars or galaxies) painted on a balloon that is being inflated. All the stars appear to be rushing away from one another. Edwin Hubble is often credited with being the first to recognize this fact in a paper he presented in 1929. Hubble was aware that the spectra of stars appeared to be slightly, or in some cases much more than slightly, shifted toward the red end of the spectrum, although he remained cautious for some time concerning speculations about exactly what this might mean.

The red shift that Hubble, and everyone since has observed, occurs due to the unusual properties of light. If two cars crash into each other, each going 30 MPH, any school child can tell you that it would be the same as a single car hitting a stationary vehicle while going 60 MPH. The velocities are added, which is why head-on collisions are so terrible. In 1887, Albert Michelson and Edward Morley preformed an experiment in which they measured the speed of light in relation to the movement of the Earth. The speed of light was known at the time and it was assumed that, like the colliding cars, light would appear to move faster if the Earth were moving toward the light source rather than at a right angle to it. To the experimenters' shock, light seemed to go the same speed no matter if one were racing toward it or not!

Einstein's theory of relativity eventually helped to explain the unusual findings. Without attempting to get too technical, Einstein showed that the speed of light was a constant (unlike a variable, such as the outside temperature, which is free to vary). Light will never change its speed through a vacuum. When one races toward it, rather than appearing to impact the viewer at greater than light speed, the light constricts by shifting its frequency toward a more energetic blue, and thereby shows an increase in energy which accounts for the combined motions of the observer and the light without ever causing the speed of the light as measured by the observer to change. Similarly, if a light source is rushing away from an observer, it will still appear to arrive at the constant speed of light, but be of a lesser energy and be stretched to a longer, or redder, wavelength. It is interesting that the energy of a sound wave is tied to its amplitude...the louder it is, the more energy it has. Its frequency, or pitch, is unrelated to the total amount of energy. Light, on the other hand, is more powerful at higher frequencies! A very dim blue light will hit a target with a lot more punch than a bright red light. This is why X-rays, which are nothing more than very high-frequency light waves, have the power to penetrate.

The point is that stars in other galaxies appear to be red-shifted, meaning that they are all moving away from us. The universe is expanding. By tracking the expansion back to its point of origin, it is possible to deduce that the universe started out as a far smaller and compact entity, and expanded outward. This outward expansion was so forceful as to garner the name the "Big Bang." Since the Big Bang was first proposed, support for it has been forthcoming from a number of important observations. First came the discovery in the 1960s of the afterglow of the explosion (in the form of background microwave radiation coming from all parts of the sky). Since then many more observations have supported the existence of the Big Bang. One of the most prominent to examine the afterglow of this immense explosion is the Wilkinson Microwave Anisotropy Probe (WMAP) which was launched by NASA in 2001 and which has yielded the best estimate so far obtained of the age of our universe—13.7 billion years.

Just before the universe exploded into what we see today it was smaller than the period at the end of this sentence. What is often hardest for people to imagine is that when the universe exploded, it didn't explode into a surrounding space. If, in your mind's eye, you imagine the Big Bang from a view outside it, the way you might imagine watching any explosion from a distance, suffice it to say that there wasn't any place at the time from which you could have watched. Space itself was created in the explosion. Before space came into existence there was nothing surrounding the tiny universe.

It might at first appear that "space" and "nothing" are the same things. But that isn't true. Think of it this way: Between the Moon and Earth is space. If space were the same as "nothing," then there would be "nothing" between the Moon and Earth. And, if there were nothing between the Moon and Earth, they would be touching. But they aren't touching. *Something* separates them. That "something" is space. Space is a thing.

According to Einstein's theory of relativity, neither matter nor energy can ever surpass the speed of light. Light itself can travel at the speed of light (which figures), but not surpass it. Matter can approach the speed of light, but not obtain it (at light speed, any bit of matter, no matter how small, would have an infinite mass). So when the Big Bang occurred, the light barrier limited how fast matter and energy could be blown outward. Since the universe is 13.7 billion years old, this would mean that no object could possibly be more than 13.7 billion light years away from us. Indeed, the most distant objects yet observed appear by their red shifts to be about 13 billion light years away. But space itself is an exception. There is nothing in relativity theory that prohibits *space* from expanding far faster than the speed of light.[*]

---

[*] Interestingly, Einstein's theory only forbids particles from surpassing the speed of light, but not from maintaining super-light speeds. With this in mind, physicists have looked for particles that forever move at super-light speeds, never slowing down below the light barrier. However, no such *tachyons* have ever been observed.

Keep that odd quality of space in mind while we consider another very strange thing about space...it can curve. It all depends on how much matter there is in it. We shall discuss why it curves a bit later, but suffice it to say for now that it can curve either positively or negatively. The question then becomes, how curved might the universe be? If it is curved positively (or as the cosmologists say, Omega is positive) the universe will curve back on itself. To simplify it greatly, that would mean that if you started out into space and never made a turn, you'd eventually end up about back where you started. Magellan realized that the surface of the Earth was positively curved which allowed him to circumnavigate it. (Well, okay, not he himself, since he was speared by natives of the Philippines while forcibly attempting to convince them that his religion was better than theirs...but a few folks on his ship made the full round trip).

If Omega is negative it gets a bit harder to imagine. Basically, space would be saddle-shaped in its curvature and the farther out into space you ventured (again without making any turns) the farther you would get from *everywhere* (anyone who has ever missed a turn-off on the freeway knows the feeling—although I admit that that wouldn't actually be an example of negative space curvature).

Since Omega has an apparently large range of possibilities as to what it might actually be, it would really be odd if Omega turned out to be exactly zero. But zero appears to be *exactly* what it is (or within .1 of zero, which is as close as anyone has been able to measure it). This would mean that space isn't curved...it's flat...flat as a two-cent pancake. Once again we are faced with what at first glance appears to be an amazingly unlikely coincidence...of all the numbers it might have been, it turns out to be exactly zero or fantastically close to it...what were the odds? But rather than shake our heads, we might turn for an answer to the fact that space can expand at super-light speed.

Imagine if the Earth weren't 8,000 miles in diameter, but rather billions of light years in diameter. We'd probably still think it was flat today. If space is big enough, and I mean really big, it might be curved in some fashion, but still appear totally flat to our measurements. Just how big might space then be?

We know that space has a diameter of at least 13.7 billion light years because we can see about that far. But because space could have expanded at super-light speeds, and might still be expanding, it could be much larger than that. As I said, it could be so large that it might *appear* to us to be totally flat even if it were in fact quite positively or negatively curved.

In an effort to understand just how big the actual universe might be (as opposed to the part of the universe we can observe), Russian cosmologist Andrei Linde, now a professor at Stanford, calculated that the actual size of the universe might currently be ten to the tenth power *to the twelfth power.* That's ten to the tenth to the twelfth! Or, if you like, that's a one followed by a trillion zeros! I would write that number out for you, so you could better appreciate its size, but it would require that this book be about 400 million pages long and my editor raised an eyebrow when I suggested doing it.

In his book, *The Whole Shebang,* author Timothy Ferris describes an instance when Professor Linde showed him that amazing calculation for the diameter of the actual universe. Ferris said:

> I asked one of the stupidest questions of my life: 'In what units is that [number] expressed ?....I mean, is it centimeters? Light-years? Hubble radii?'

I hope that Ferris didn't think it was really that stupid a question, because I must confess to wondering the same thing when I discovered that the diameter of the actual universe might be a 1 followed by a trillion zeros. But the fact is, with a number *that* big, the units just don't matter! Think of it this way, if one followed by a trillion zeros is expressed in light-years; it can be also expressed in millimeters as a one followed by a trillion and eighteen zeros. As you can see, the difference between 1,000,000,000,000 and 1,000,000,000,018 is hardly worth talking about. Any way you look at it, it is a staggering number—so huge that no one really cares about the units of measure. It is mind numbing to realize that the addition of each one of the trillion zeros denotes a universe 10 times larger than the one before. Ten times bigger, ten times bigger, ten times bigger—a trillion times.

Apparently, the known universe with its 13.7 billion light-year diameter might be just an itty-bitty, teeny-weeny, micro-speck contained within a humongous, titanic, monstrously gargantuan SPACE. The question now becomes, "What's in all that extra space anyway?" At first glance it might seem reasonable to assume that it was—well—empty. Admittedly, that would be an awful lot of empty, but what else could be in there?

Suppose you could travel to the edge of the known universe and look beyond the horizon into that "extra" space? Cosmologists actually refer to it as a horizon because it is as far as the speed of light will allow us to look. In fact, if you could instantly travel to the edge of the universe and look into that massive space, you would only be able to look to the next horizon (another 13.7 billion light-years away) at which point you would be prevented from seeing further by yet another horizon. Each horizon is set by the age of the universe, which limits how far light could have traveled during that amount of time. But suppose you could look beyond the first horizon, into at least a little bit of that massive space? What would you see if anything?

In Arthur C. Clarke's *2010 Odyssey Two*, we discover that the last words heard from astronaut Dave Bowman just before he disappeared were a cryptic, "My god, it's full of stars." Although it is complete speculation, I have a feeling that those would be the words anyone might choose if he or she had a chance to gaze into that massive, endless space that exists beyond the horizon. Quite a number of cosmologists, astronomers, and physicists are beginning to believe that the stars we can see, the ones that you recall equal the number of grains of sand on all the beaches and deserts on Earth, are a mere nothing when compared with the number of stars that are really out there—that is, out beyond the many horizons past which our vision cannot extend.

ZERO SUM GAME

With sturdy shoulders, space stands opposing all its weight to nothingness.
Where space is, there is being.
—Friedrich Nietzsche, "Thus Spake Zarathustra"

If this is true, then the numbers of stars that actually exist, and the number of planets that orbit them, are numbers so huge, so mammoth in size, that just the thought of such a number would be enough to make any cosmologist swallow his gum. In all practicality, we could just about warrant that there would be so many Earth-like planets out there with sentient beings in residence, that the odds of your consciousness coming up again would be just about guaranteed. That is, assuming that whatever is responsible for your consciousness could be repeated. If it could, well, such a universe would just about make it a certainty that it would.

But from where would all these other stars come? In fact, from where do the stars *that we can* see originate? How could all that have come from nothing? Amazingly, the current arguments put forth by cosmologists and physicists studying the problem are that they didn't all come from nothing, but rather that *they still are nothing*. At first glance this argument might seem to be a bit obscure, since the universe clearly has stuff in it, stuff anyone can plainly see by merely looking up on any starry night, or for that matter, by taking notice of the sun or the earth during the day. The entire universe obviously isn't "nothing;" it clearly is something. But, in terms of the physics and mathematics that might not exactly be true.

This might be a good time to consider the law of physics known as the conservation of matter and energy. This law basically states that matter and energy can neither be created nor destroyed. If a piece of wood is burned, the total of its mass and energy will be conserved, inasmuch as you could measure the ash remaining after it was burned, the vapor and smoke given off, the heat, light, and sound generated, and discover that the "wood" was all still there after it was burned, just now in different forms of matter and energy. None of its original mass or energy would have been destroyed (i.e., removed from our universe). So, if neither mass nor energy can be destroyed *or created*, how could all these stars and things have appeared in the first place? The answer lies with gravity. Gravity, as it turns out, is a negative energy. Allow me to take a little side-step at this point and take note of the fact that energy and matter are one in the same; we only use the different words "energy" and "matter" to describe this "thing" when it seems to us to appear in a significantly different form. In fact, until Einstein showed that they really were the same with his famous formula $E = MC^2$ [(E)nergy equals (M)ass times the (C)onstant speed of light squared)], people assumed that matter and energy were fundamentally different things rather than the way we see them today, as manifestations of the same "thing." Gravity as a "negative" energy can then counterbalance "positive" matter in equations so that the whole universe actually adds up to zero. In this way, matter has not been "created." The universe actually still is nothing!

At about this point you might be thinking that this sounds like something lawyers would come up with rather than physicists and I can't blame you. For this reason it is worth taking a quick look into the logic of this whole business so that you might better appreciate it.

For help I will turn to the work of physicist Alan Guth, whose superb book *The Inflationary Universe* delves into this issue in far greater detail than I shall here. In his book, Guth uses Newtonian mechanics to show that gravitational energy is negative, although he readily admits that Einstein's theory of relativity will lead to the same conclusion.

To follow this logic we will conduct what scientists call a thought problem, a problem that is conducted in our imaginations. Begin by imagining a bizarre planet that has a thin crust and is completely hollow inside; basically, a hollow sphere. I know that this would be a pretty odd world, but since it is only here to serve our thought problem, let's accept that there is such a world.

Like all worlds, this world produces a gravitational attraction (see Figure 11). Granted, the gravitational attraction would be greater if the planet weren't hollow, but even a hollow planet will produce some gravitational pull owing to the mass of its crust.

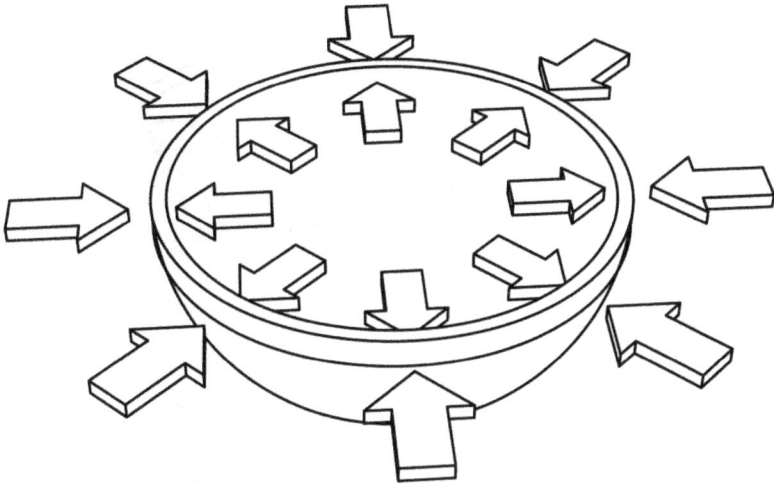

Figure 11. Like all masses, this hollow sphere of a planet creates a gravitational pull toward its surface from both within and without the planet. The arrows represent the direction of the gravitational pull. Its pull, however, is less than if it were a solid sphere.

If you are anywhere near this planet you will be able to measure its downward pull, and the closer you get to the surface, the stronger the pull will be, with the strongest gravitational force being right on the surface. The same is true of non-hollow spheres such as the Earth or other planets;

the strongest pull is right at the surface.* Gravitational attraction would emanate from this hollow sphere outward into all the space that surrounds it. Even at very great distances from this planet, there will be an attraction, although at large distances it would be greatly diminished. Again, this is true for all worlds, solid or hollow.

The gravitational force inside the sphere is more interesting. At any point within the shell, attraction will be greatest along the route that is the shortest distance to the crust (see Figure 12).

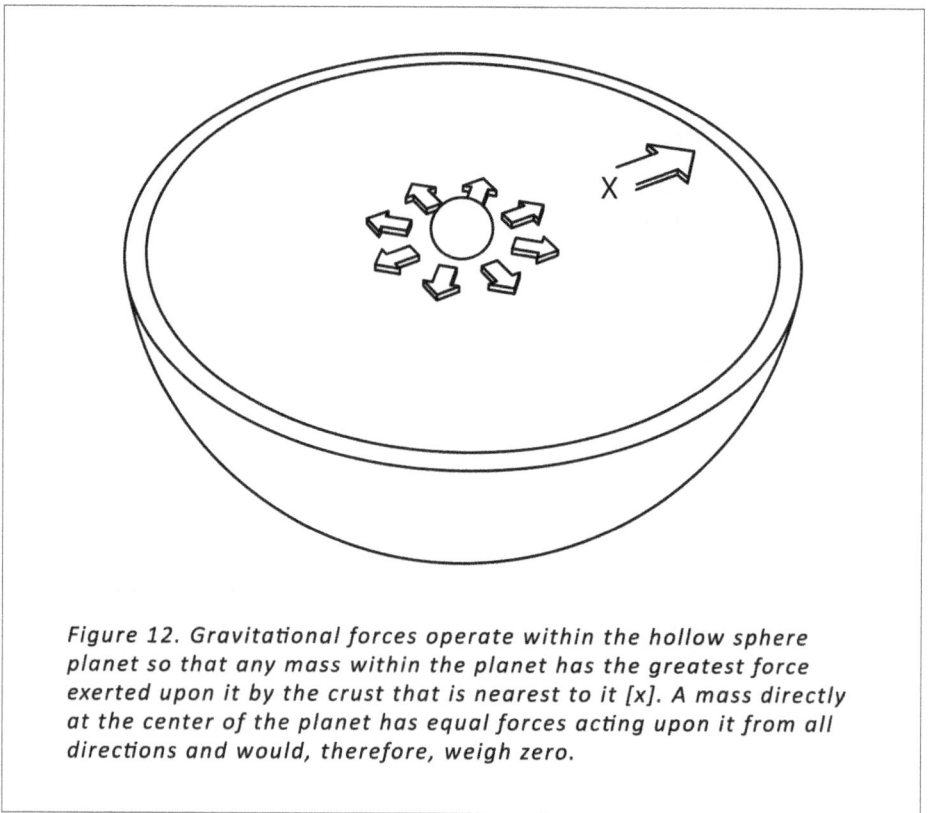

Figure 12. Gravitational forces operate within the hollow sphere planet so that any mass within the planet has the greatest force exerted upon it by the crust that is nearest to it [x]. A mass directly at the center of the planet has equal forces acting upon it from all directions and would, therefore, weigh zero.

---

* From time to time, enterprising physics teachers will take their classes to the top of the Empire State Building in New York, where sensitive instruments can clearly show that the pull of gravity is less than it is down at street level.

Interestingly, if one adds up all the forces acting upon all the possible points within the sphere, the forces will cancel each other out and equal zero. This can be represented by a single point that is equidistant from all points on the surface, a point that would be right in the center of the sphere. Anyone residing at that center point would be weightless as the gravitational force there is zero. Indeed, anyone at the exact center of the Earth would likewise be weightless. We can therefore say that there is a gravitational field that adds up to zero *within* the sphere, while there is a gravitational field that adds up to more than zero *without* the sphere. Another way of expressing this is to say that the gravitational field inside of the sphere is zero, while outside it is greater than zero.

Now, let's really get out imaginations going. Imagine thousands of huge electric generators, extremely massive ones that would take an enormous amount of energy to budge, and imagine that they are all in geosynchronous orbit above our spherical planet (an orbit that keeps them in one location because they make each orbit in synchrony with a single rotation of the planet, like TV broadcast satellites do). From each generator runs a super-thin, but very strong line, which is anchored firmly to the planet's crust. If we pull on any of these lines it will turn a wheel in the generator to which it is attached, thereby making electricity that we can then beam off to other worlds that might need it.

Now, let's compress our spherical planet equally from all sides so that it becomes a new hollow sphere, but with about half of its original diameter. It is a smaller world now, its crust is thicker (since we have to conserve the matter in the crust as it was compressed), and its center is still hollow. However, when we collapsed the surface of our world inward we also pulled equally on every line attached to every generator. The generators hardly moved at all since they were so massive, but rather the lines spun the wheels in the generators which, in turn, made lots of electricity which was beamed off (as microwave energy, if you like) to other needy worlds nearby. We have now extracted a bunch of energy from our local system.

But look at the aftermath of all this energy extraction (see Figure 13) and you will see that there is now a shaded area of space that is different.

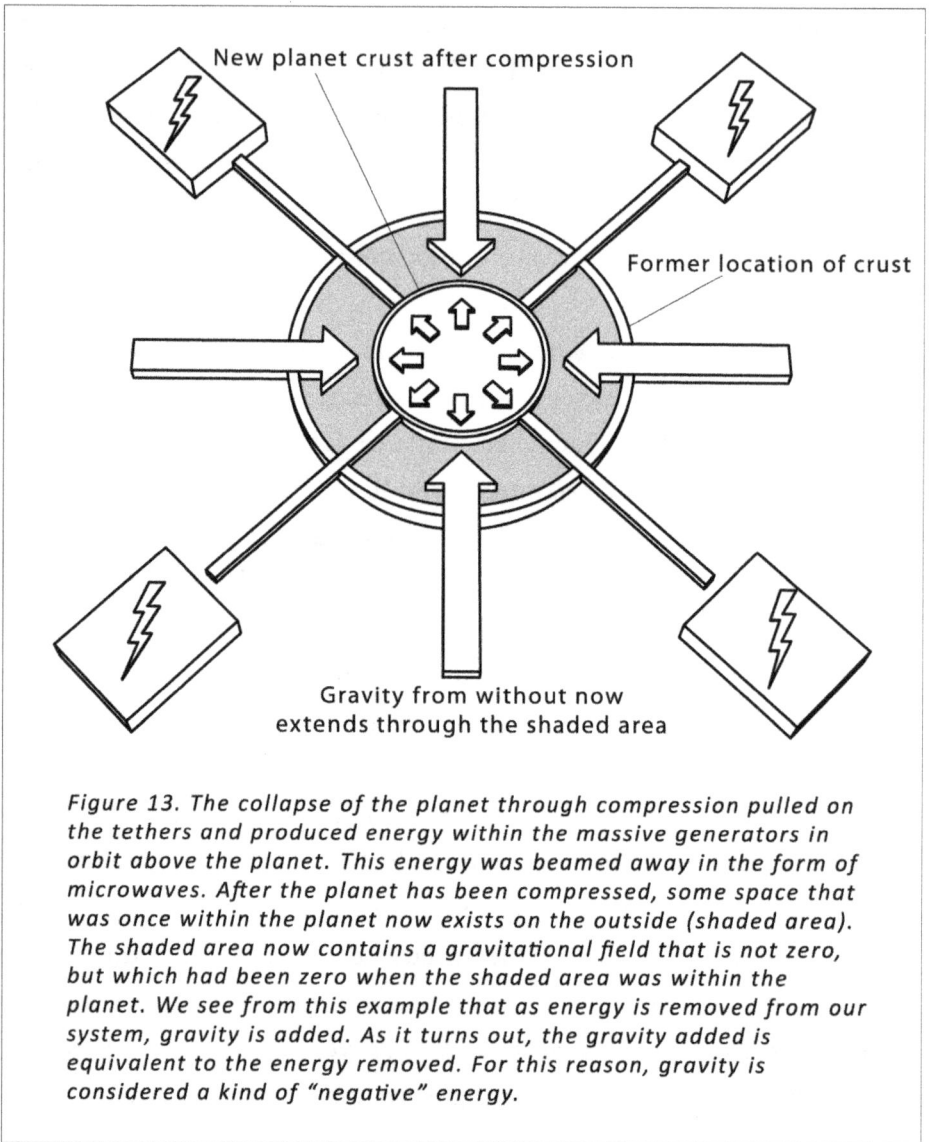

New planet crust after compression

Former location of crust

Gravity from without now extends through the shaded area

Figure 13. The collapse of the planet through compression pulled on the tethers and produced energy within the massive generators in orbit above the planet. This energy was beamed away in the form of microwaves. After the planet has been compressed, some space that was once within the planet now exists on the outside (shaded area). The shaded area now contains a gravitational field that is not zero, but which had been zero when the shaded area was within the planet. We see from this example that as energy is removed from our system, gravity is added. As it turns out, the gravity added is equivalent to the energy removed. For this reason, gravity is considered a kind of "negative" energy.

The shaded area used to be inside of the sphere, but since the sphere was collapsed, this area is now on the outside. The shaded area is also different because it now has a gravitational field associated with it that is greater than zero when before it had a field that was equal to zero. Another way to say this is that *as our system gave up energy, it acquired gravity*. Gravity, therefore, represents *energy that is gone from our system*. Gravity balances the missing energy so as to keep things equal. Since energy is a form of matter, gravity can also function to balance out matter on the cosmic scales so as to keep things equaling zero. Amazingly, matter that just seems to appear is not "created" if a counterbalancing gravitational force also appears at the same time, because *together they add up to zero*. Nothing added—nothing created.

I mention all of this because it appears that following the Big Bang and the stupendous expansion of space that followed, a "froth" of matter and energy counterbalanced by the appearance of gravitational force, filled the universe. Mathematically speaking, nothing was "created" when this occurred because it all added up to zero.

At this point you might well ask, "Why would there be some 'froth' that would coalesce into matter and energy, and eventually galaxies—why wouldn't there have been simply nothing at all?" Well, that's a good question. Quite honestly, no one knows, but there has been speculation both by philosophers and physicists. Let's take a look at the philosophical view—and I suggest that you read the next paragraph very slowly.

Philosophically speaking, one might argue that *nothing*, quite literally *no thing*, can only be defined by the absence of *things*. Nothing presupposes the existence of something that is no longer there. If there is *absolute* nothingness (i.e., there was never anything), then there couldn't even have been a thing that is now gone. So nothing couldn't actually be nothing since it would now have to be defined without the possibility of there ever having been a thing, so nothing itself couldn't be, since it is understood to be the absence of things. Therefore, nothing can't be *no thing* since there wouldn't be any *thing* for which it could be *no* about. Logically, *absolute nothing would be an unstable condition.*

To a physicist, the forgoing would most likely be viewed as a lot of gobbledygook. He or she would more likely be interested in well-defined theories and solid evidence that might support such theories. Taking this more empirical approach physicists have come to the conclusion that, well, *absolute nothing would be an unstable condition.*

Modern cosmologists refer to this instability as a *vacuum fluctuation.* In the 1960s, Edward P. Tryon, an assistant professor at Columbia University, became the first to suggest that the entire universe might be a vacuum fluctuation. Alan Guth describes Tryon's idea in *The Inflationary Universe*:

> By a *vacuum fluctuation*, Tryon was referring to the very complicated picture of the vacuum, or empty space, that emerges from relativistic quantum theory. The hallmark of quantum theory, developed to describe the behavior of atoms, is the probabilistic nature of its predictions. It is impossible, even in principle, to predict the behavior of any one atom, [this is known as the Heisenberg Uncertainty Principle] although it is possible to predict the average properties of a large collection of atoms. The vacuum, like any physical system, is subject to these quantum uncertainties. Roughly speaking, *anything* can happen in the vacuum, although the probability of [something as complex as] a digital watch materializing is absurdly small. Tryon was advancing the outlandish proposal that the entire universe materialized in this fashion!....[Tryon] had understood the crucial point; the vast cosmos that we see around us could have originated as a vacuum fluctuation—essentially from nothing at all—because the large positive energy of the masses in the universe can be counterbalanced by a corresponding amount of negative energy in the form of the gravitational field. "In my model," Tryon wrote, "I assume that our universe did indeed appear from nowhere about 10^{10} years ago. Contrary to popular belief, such an event need not have violated any of the conventional laws of physics."

So, the next time that you are looking up at the sky, or anything else for that matter, and someone asks, "What are you looking at?" you can quite rightly say, "Nothing."

The *known* universe with its septillions of stars and planets is, as we have noted, amazingly vast. Imagine that these two dots (. .) represent the sun and the star nearest to it (Proxima Centauri). It would be an amazing accomplishment if human beings were ever able to make a journey to Proxima Centauri and return home. But when we use the scale of the two dots on this page, the idea of traveling to the furthest known star becomes just a bit daunting, because it would be about 10,000 miles away!

Theoretical research by physicists such as Steven Weinberg of Harvard and Soviet physicist Yakov Borisovich Zel'dovich suggests that matter and energy might materialize from expanding space if the space were expanding fast enough. This has led to speculation that the monstrously sized *real* universe, (all that might be beyond the light-barrier horizon, past which we cannot see, out to perhaps a 1 followed by a trillion zeros or perhaps even more), could well be filled with countless galaxies (*zillions* is the amount that comes to mind—although that isn't a real number, it just seems appropriate to use it in this instance).

It has also been recently discovered that our expanding universe is expanding at an *accelerating rate,* which is the opposite of what one might expect. This has led cosmologists to search for a repulsive force that can counter gravity, a force some have called dark energy. The known universe, therefore, is becoming ever larger at an accelerated rate.

The stunning size of what might be the real universe reminds me of when I first arrived in Tempe, Arizona, to attend graduate school. It was 117 degrees in the shade as I pulled into town in my Volkswagen. As I waited in the sizzling heat for a light to change, a man in a car next to me, whom I am guessing saw my out-of-state license plate, actually shouted over at me, "Hot enough for you?" He actually said that, as though I was well known in Arizona as someone who had been advocating higher temperatures for years and could now be blamed for the current scorcher. How could I answer him? I thought of saying, "Actually, it was hot enough for me 40 degrees ago," but that didn't seem to be a particularly witty reply. So I just dumbly nodded. Then the light changed and I never saw him again.

I tell you all this because now that I contemplate the size of what is most likely the real universe, I strangely find myself identifying with that man in Tempe, and am oddly compelled to ask, "Big enough for you?" But just in case you're not totally overwhelmed by the thought of a universe filled with stars and planets that has a diameter of a 1 followed by a trillion zeros, I shall fall back on my experiences viewing infomercials and shout, "But wait, there's more!"

THE MULTIVERSE

As an individual, I myself feel impelled to fancy ... a limitless
succession of Universes.... Each exists, apart and independently,
in the bosom of its proper and particular God.
                                        —*Edgar Allan Poe, "Eureka"*

For too many cosmologists it is not correct to consider the idea of a "known universe" vs. a "real universe" since there is no way that we could ever see significantly beyond the nearest horizon 13.7 billion light years away, at least not for a very long time. Instead, they prefer to think of any stars, galaxies, or worlds that are beyond the horizon as properly belonging to *another universe.* Each additional horizon in our trek outward to a distance of a 1 followed by a trillion zeros would mark the border of yet still another universe. The most often used analogy is to imagine a child blowing soap bubbles from a bubble pipe. The full mass continues to grow and expand, but each bubble is self-contained; its own little universe. Many cosmologists like to think of the perhaps endless number of horizons defining each universe as "an eternity of bubbles." The late Carl Sagan said in his book *Pale Blue Dot,* that he preferred to refer to our universe as *the universe,* and the whole works, if it existed, as *the cosmos.* More commonly, cosmologists have come to refer to the "whole works" as *the multiverse.* Modern inflationary theories, which help to explain the accelerating expansion of our universe, demand the existence of multiple universes. The idea of multiple universes is no longer considered bizarre by physicists and cosmologists. Today, because of the observations that the expansion of the universe is accelerating, the idea has become fully imbedded in the mainstream and well accepted.

The thought that there might be many more stars and galaxies beyond the distant light-barrier horizon is one of many possibilities, and perhaps not the most likely one. Modern inflation theory predicts the existence of other universes in yet another way. The conditions that lead to the *false vacuum* that appears to have produced our own universe, that is the quantum fluctuations in a vacuum that led to the Big Bang, might not be all that rare an occurrence. As space expands, inflationary theory predicts that an infinite number of big bangs will occur, yielding new "eternity of bubbles" universes that completely break off from our own. This differs from the view that other universes are just beyond the light-barrier horizon inasmuch as these individual bubble universes are almost never likely to come in contact with one another. In fact, entire other universes might exist less than a millimeter away from you. Think of them in this way. They are all around us all the time, but slightly out of tune with ours, much the same as there are radio stations broadcasting about you all the time, but you are only aware of the one to which your radio is tuned. In the same way, an endless number of other universes might be all around us, but out of phase with us.

It is quite possible that other universes are being created all the time. In fact, there is even some concern that someday during an experiment here on Earth someone might create the conditions that lead to a false vacuum and inadvertently make a new universe in his or her laboratory. Calculations indicate that this would cause an explosion equal to a large hydrogen bomb (a little bang as far as cosmological blasts go), and that the new universe would break off from ours never to be seen again. While certainly this would be an interesting result in anyone's lab experiment, such an outcome would likely put a crimp in the weekend of whoever conducted it. Some physicists, such as Michio Kaku, have even speculated that very advanced civilizations might be able to leave a dying universe (one that has expanded so much that it was in the process of cold death) by creating a new universe and finding a way to enter it just as it is created, thereby leaving the laboratory forever along with the newly created universe. As a general rule, though, I wouldn't suggest trying it, at least not anytime soon.

Universes created in this fashion can be thought of as parallel universes to our own. Modern theoretical physics (in particular, string theory) indicates that such universes might exist on great "membranes" that under certain conditions could interact with our own universe, but which mostly would remain independent from us. In his book, *The Universe in a Nutshell*, Stephen Hawking argues that there is reason to believe that gravitational waves might have the property of passing from one membrane to another, something which might be possible to measure and, indirectly, someday yield evidence for a parallel universe.

As you recall, we discussed the possibility that the real universe (as opposed to the observable universe) might be 10^{10} to the 12th in diameter. But, believe it or not, there are some cosmologists who think that is way too small! Modern theory predicts that the creation of new universes is a never ending process.[*] This could either mean one titanic universe that never stops expanding (an eternity of bubbles—also known as the conservative universe option—and who knows for how long that's been expanding, since our visible universe might be a recent addition to the mass of bubbles), or it could also mean many parallel universes that break off from each other, or perhaps both.

The possibility of many universes also helps to answer an important question first suggested by the French writer Michel de Montaigne and

---

[*] It should be pointed out that modern physics theory is strongly supported by vast amounts of solid evidence gained from decades of experiments with particle accelerators. We are not just talking about some guess somebody has pulled out of a hat. In fact, the standard model has been found to be correct in its predictions, when its predictions can be tested, down to an error of one in 10 billion.

later elaborated upon by American astronomer Carl Sagan. In a nutshell, the question asked is why Newton's law of gravity should have the inverse square relationship $1/r^2$ where r equals the distance between the centers of two masses. Why wasn't it $1/r^3$ or something else? Of course, if it had been something else, the Earth would not have formed and he wouldn't have been alive to pose the question. But still, it is a fundamental question, namely, why are the laws of physics what they are and, as a further extension of the question, where did they come from?

Carl Sagan argued that the existence of many universes might provide an answer to this profound question. If the laws of physics appear randomly, then there would be many universes that had different laws, universes that were not only very strange from our perspective, but totally inhospitable to the formation of life. We live in a "baby bear" universe, one that is just right, because after dealing enough hands of poker a royal flush finally got dealt. There is nothing magical about it. In this view, the laws of physics *can be anything*; it's just that in our universe we got a lucky mix that led to our world and to us. It might even be possible for the laws of physics to be different for each bubble within a single infinite universe, further emphasizing that each bubble really could be considered a universe unto itself. If there are an infinite number of bubbles, though, there would still be an infinite number of stars and planets out there as there would also be an endless number of universes with laws like ours despite there being an even greater number of universes (bubbles) with different laws.

While modern cosmology is a fascinating area, you might be wondering how consideration of it helps us to address the great question, "What happens to us after we die?" I think it does help, because we can now see why we live on a "baby bear" planet. You will recall earlier that we were wondering how it could possibly be that there was this wonderful planet called Earth, that was not too close or too far from the sun, that was just the right mass, that had a moon to stabilize its tilt, that was in a solar system with a giant planet (Jupiter) that could sweep up most of the deadly asteroids floating about, and that had lots of water, as well as a lot of other life-supporting goodies. We also wondered about the formation of life and what might be the odds that one branch of the evolutionary tree might lead to human beings. (Despite how you might feel, we are not at some pinnacle of the animal kingdom, but rather we are just one of many off-shoots the kingdom might just have easily done without).

We also asked, "What were the odds?" Perhaps now we can understand how such a perfect planet came to be. It is simply that a vigintillion-to-one-shot is no big deal if there are a billion-trillion-vigintillion chances for it to happen! I play poker on occasion, but I have never had a natural royal flush dealt to me ("natural" meaning that no wild cards were used). However, if a trillion hands of 5-card poker were dealt to me I could expect, on average, to receive about a million and a half natural royal flushes before the draw (in other words, the first five cards dealt to me would be the royal flush). In fact, the odds of *never* getting one after a trillion hands would be astronomical. At least one royal flush in a trillion hands of poker is as close to a sure thing as you are ever likely to find. Therefore, if we consider a monstrously sized universe filled with stars and planets, and then consider a nearly infinite number of such universes, some of them surely with a "baby bear" set of physical laws, is it any wonder then that our Earth came to be? In fact, the odds become so certain for the existence of a fantastically huge number of Earth-like worlds that one might seriously entertain the prospect that there would even be Earths nearly identical to ours, perhaps with only the tiniest of differences, such as their Cleveland, Ohio having a Maple Street where our Cleveland, Ohio has an Elm Street. These sorts of things begin to happen when one plays about with infinity.

If we imagine our universe (without even bothering to refer to parallel universes) to be larger than professor Linde's calculations suggest it might be, but which modern physics suggests it could be, we can even make some fun calculations about what's beyond the 13.7 billion light year horizon of our universe. In *The Infinite Book*, John Barrow, Professor of Mathematics at Cambridge and Fellow of the Royal Society has done just that. He knew that it was roughly 10^{27} meters to the edge of the universe. With that in mind, as well as the known distribution of planets and stars, he roughly calculated that you would have to travel $10^{10^{28}}$ meters (way the hell and gone past the horizon and a number so huge that it hardly matters if you call it meters or anything else) before the odds were that you would find an exact copy of yourself; $10^{10^{50}}$ meters before you found an exact copy of Earth; and $10^{10^{119}}$ meters before you found an exact copy of our own little universe in one of the endless "bubbles." If Linde is considering the universe to only be a whopping $10^{10^{12}}$ it actually might be too small to harbor an exact copy of you. But then again, that wouldn't be at all true if there are an infinite number of other universes in parallel with ours.

Of course, if there are that many life-hospitable planets in this universe and others, there will be that many more chances for your consciousness to reoccur. The basic idea is straightforward. We are each individuals, each with a different consciousness. The conscious mind appears to be "of the body" rather than some mysterious spirit that can never be measured or defined. We are different from one another because each person's conscious mind is different in some way, in the same way that fingerprints are different. You are you, and I am I, because the combinations of brain chemistry that led to you or me are peculiar to each of us. If there are ample chances for the combinations that led to your consciousness or mine to reoccur, we will experience life once again following our deaths. Because it is most likely the case that our observable universe is filled with planets that could support intelligent life, and because there might well be nearly an infinite number of universes, with an infinite number of worlds we can assume that the chances for our "consciousness combinations" reoccurring become nearly infinite, and therefore assured. As mathematicians say, "In a universe of infinite size, any non-zero probability event must occur infinitely."

So there it is...all wrapped up and tied with a pink ribbon. There probably is no "death" as most imagine it. There is only life after life as the combination that leads to your consciousness keeps coming up again and again all over this universe, or others. The whole concept is very straightforward and logically appealing. There is only one problem with it...it doesn't work.

THE SUPERVENIENCE PARADOX

When a shadow flits across the landscape of the soul where is the substance?
—Henry David Thoreau

Just when things were going along so nicely, a great big greasy monkey wrench gets tossed right into our gears...the supervenience paradox. The philosopher Donald Davidson introduced the idea of supervenience into the theory of mind in 1970. Davidson said:

> [M]ental characteristics are in some sense dependent, or supervenient, on physical characteristics. Such supervenience might be taken to mean that there cannot be two events alike in all physical respects but differing in some mental respect, or that an object cannot alter in some mental respect without altering in some physical respect.

Building upon this idea, I have chosen to call the conundrum we are about to discuss the supervenience paradox, but I suppose it might just as easily be called by another name, and I am sure that it has, many of which are unfit to print. Perhaps you realized this problem earlier on. If so, I congratulate you.

The predicament can best be understood by asking a question and then by presenting a thought problem along with some examples. The question is as follows: If, after you die, your consciousness can reoccur and cause you to be alive once again, what happens if the combination that leads to your consciousness reoccurs in another body *while your first body is still alive*? Oops...I hate when that happens! Now who are you? Are you both people at the same time? In the words of Meredith Wilson (almost), "ya got trouble, folks, right here in River City...with a capital 'T' and that rhymes with 'P' and that stands for paradox."

Let's use a thought problem to better illustrate the philosophical mess into which we have now gotten ourselves. Imagine (which is a good way to begin a thought problem) that you are in a room with two chairs, one

145

chair labeled *original*, and the other labeled *copy*. I now invite you to sit in the chair labeled *original*. I do this to make the point that I will consider you to be the "original you." Next I will bring out a very special machine that has the ability to replicate objects by making exact copies of them. I believe on Captain Picard's Enterprise this sort of machine was in use (curiously only to produce food—mostly Earl Grey tea, hot).

However, before I take another step forward I realize that there will be some who, upon hearing of my plans, will cry out "Heisenberg!" Of course they would be right. Werner Heisenberg's Principle of Uncertainty clearly shows that it is not possible to ever know both a particle's exact location and its exact momentum at the same instant. I suppose because this is a thought problem that I could envision a world in which the laws of physics were different, but I think that that would be cheating inasmuch as I want to stay within the boundaries of what seems to be the universe in which we live. So I will agree that the scanner imagined by Ray Kurzweil that might conceivably be able to recreate your mind in every exact detail within a machine, would never be possible, at least not in our universe.

My plan then is to give Heisenberg a bit of a head fake so as to get around him. I will do this in the following way. As stated, there are two chairs, one labeled "original," in which you are seated, and the other labeled "copy." Now I will bring out a machine that will attempt to recreate the matter and energy that comprises you at this given moment and build a new "you" in the chair labeled copy (as best as it can within the constraints placed on us by the uncertainty principle).

The machine now begins to whirr and buzz and soon enough a ray passes over the "copy" chair and builds a person right there who looks exactly like you. (Needless to say, considering $E = MC^2$, the electric bill for making the matter used to construct your copy might be somewhat breathtaking). The Heisenberg Uncertainty Principle forbids us from *deliberately* making an exact copy of you owing to the fact that we can never know the exact location of a particle and its momentum at the same time...not to mention that after scanning you, time would pass before the replica could be made, thus allowing the original "you" to alter in the intervening time, thereby causing the copy to be a match of you some seconds earlier, and not an exact copy of you as you are now. So I am going to postulate that the machine, in an effort to make the best copy of you possible, just happens to make an exact copy of you in the "copy chair" by wild, random, chance (lucky us). Of course, in keeping with the constraints of Heisenberg, no one running the procedure will ever know that this has happened (nor could they ever know). In fact, just to be extra safe, I am going to argue that you and I would not know—I am only asking *what would it be like if such an event did occur whether or not we knew it had happened?*

So, what would happen if an exact duplicate of you, down to the finest detail, were to appear in the chair next to you? This is our thought experiment. If your experience following the creation of the copy were that you simply continued to sit in the "original chair" while what seemed to be an identical twin of yours appeared in the chair next to you (which is intuitively what I believe most of us would think would happen), a problem immediately arises. If you are still you, and the exact copy of you in the next chair is not you, but someone else, how can this be if consciousness is part of the body? Here, sitting next to you is an *exact* recreation of your brain, and yet it is someone else. Such a result would take all our previous assumptions about consciousness reoccurring due to certain brain module relationships and toss them into a cocked hat.

But suppose after the scan were completed you found yourself sitting in the copy chair amazed to see that you had somehow moved over from the original chair. What would you conclude—you must be the copy! However, this places us right back into the same mess as before, because you could now rightly ask "Who, then, is the person in the original chair?"

If it isn't you, it must be someone else, and once again consciousness appears not to be of the body, since both of you are exactly identical and yet there seem to be two different consciousnesses in play.

What else might account for different consciousnesses existing within two identical bodies? One possibility is that the two bodies are not, in fact, identical. Although we have stipulated that the scanner has, by chance alone, made an exact copy of you down to the very last atom, they are not, after all, *the same atoms.* In other words, a particular calcium atom in an arm bone in the "original you" might have an exact copy of itself in exactly the same place in the "copy you," but it would not be the *same* atom, it would be a copy of the original atom. Such a notion would imply that atoms are in some way completely individual and that no one atom could *exactly* replace another. This would mean that there could never be an exact copy of you since "you" could only be created by a particular arrangement of certain one-and-only individual atoms.

Such a possibility would explain how what appeared to be an exact duplicate of you would not have your consciousness. Even so, this idea strikes me as unlikely for two reasons.

First, there is absolutely no evidence from physics to indicate that one electron or one proton, or one calcium atom for that matter, can't directly and in all cases substitute for another electron, or proton, or calcium atom. No one has ever even seen a hint that these items have individual traits that make them each different and unique from one another. At least, we haven't seen evidence for that yet (imagine the nightmare for physicists if something like that should begin emerge as researchers plunge ever deeper into the fundamental constituents of matter).

Another reason to doubt this possibility is that it would make the odds of your consciousness having been created soar well beyond astronomical. If it not only requires a particular arrangement of atoms to yield your personal consciousness, but rather a collection of certain unique individual atoms, well, what actually are the odds? Although I can't prove it, I very much doubt that "atoms with individual personalities" is the solution to the supervenience paradox. Furthermore, as best as biologists are able to observe such goings on, it is believed that all the atoms in our bodies get replaced about every 15 years or so, by other identical atoms. Of course, there is no way to track every atom in a person's body, but if we look at the rate at which skin sloughs off, and bone is replaced, and cells actively change, it appears that after sufficient time has passed one's body has received quite an overhaul. Certain cells might last a lifetime, but the atoms of which they are comprised do appear to undergo a fairly regular exchange. This is what Ray Kurzweil meant when he said, "that fundamentally we are our 'pattern' (because our particles are always changing)." For such reasons it doesn't seem to me that consciousness could be tied to certain atoms (although who is to say that whatever is responsible for consciousness isn't made of atoms that are never replaced—or if they are replaced, their loss then causes you to become a zombie?).

So, what else is different about these supposedly exactly identical people? Well, they do occupy different space—one is seated in one chair, and the other is seated in another. Is it possible that the appearance of your consciousness requires that you occupy a particular space? That is a fascinating idea, and perhaps worth pursuing, although I am not about to make the attempt here. I am not going to try because there is no evidence from physics that altering the position of atoms in space fundamentally alters them so that they behave differently. Certainly such movements in space might subject atoms to changes in gravitational fields and such, but there is nothing about changing location in space, per se, that will cause an atom, or a collection of atoms, to alter its fundamental character.

Perhaps we need to switch gears. So far we have assumed that the copy *would* be another person—but suppose it isn't? Suppose your consciousness is recreated in the copy and, *at the same time*, remains in the original you?

What then? We could imagine a double consciousness in much the same way as we could imagine a double television image created by superimposing the output from two cameras placed in different locations. I am not talking about a split-screen effect, but rather both images simultaneously filling the full screen, and transparent to one another. As each camera went off in a different direction, you would see both overlapping images in motion at the same time. That would be somewhat confusing, to be sure, but such an outcome might solve the supervenience paradox—two bodies; one shared consciousness. The problem with this solution, however, is that it violates the constraints set upon us by the speed of light.

How would such a solution violate the light barrier? Well, let's work a quick thought problem within the thought problem on which we are currently working (I really do apologize for taking you a layer deeper into this onion).

Imagine that we send the copy you off to Mars (with your copy's permission, of course.) The original you gets to stay here on Earth. Remember, we are assuming that you are both sharing one consciousness.

Now, once the copy "you" arrives on Mars we will conduct a simple test. At a predetermined time (let's say noon G.M.T.) a colored disk (yellow) will be shown to the copy you. At the same instant back on Earth, a different colored disk (red) will be shown to the original you. Since you both share the same consciousness, what will *you* consciously experience at high noon? Remember, when Earth and Mars are closest to each other, it still takes about 10 minutes for messages (traveling at the speed of light) to go from one planet to the other. I think that now you are beginning to appreciate the problem.

If our test leads to the result that you see a red disk, and then 10 minutes later a yellow disk is added to your conscious experience, we can then say that the original you is somehow dominant, and that means that there is a difference between the two "yous." If a yellow disk is what you experience first, and then 10 minutes later a red disk is added to your conscious experience, we can say that the copy you is somehow dominant and once again we have an unacceptable difference between the two supposedly identical "yous." If we split the difference, we end up with neither of you being able to experience the sight of a colored disk that is right in front of your collective faces for 5 minutes, and then you see both disks at once. This would mean that you were both unexplainably blind for 5 minutes and that your shared consciousness was amazingly suspended in outer space halfway between Earth and Mars! What would seem to be the intuitive answer to our test, namely that you would see *both* red and yellow the moment that they were shown to "yous," violates the limit placed on us by the speed of light because the information could not be shared so quickly over so great a distance. Shared consciousness, as you can see, appears to be a dead issue unless we would like to rewrite the laws of physics.

WHERE'S EINSTEIN WHEN YOU NEED HIM?

I think and think for months and years.
Ninety-nine times, the conclusion is false.
The hundredth time I am right.

— *Albert Einstein*

So now we're stuck—with a capital S, and that rhymes with mess, which stands for the supervenience paradox. We have an original you and a copy you, and they can't share one consciousness, which makes them different, but they can't be different because they are the same.

Perhaps the solution is to suggest that when you create the copy you it will be a zombie because the original you already exists? No, that won't work. If you are conscious, and the copy you is an exact duplicate of the original you, it has to be conscious, too. Wow—where's Einstein when you need him?...Actually, Einstein might be able to help us create exactly the frame of mind that we will need to overcome the supervenience paradox.

Let's begin our effort along this line with *the principle of equivalence*, a principle Einstein used in his general theory of relativity. The principle of equivalence is a fairly straightforward idea. It states, basically, that if there is no possible way to tell the difference between two things, the two things are, in fact, *the same thing*.

In our day-to-day lives we often encounter things that appear to be quite different from one another, only to discover upon further examination that they are really the same thing, but perhaps viewed from a different angle or thought of in a different way. Matter and energy are a good example. For many centuries, matter and energy were considered to be unrelated to one another and to be clearly separate entities. Now we understand them as different manifestations of the same "thing," whatever that "thing" might be. As you recall, $E=MC^2$. In this sense, matter can be understood as a form of "frozen" energy.

Sometimes people will have heated arguments about whether something is really one thing or another, when in fact what they are arguing about is the same "thing." Perhaps you have heard of the amazing duality of light—sometimes a photon of light acts like a wave, and sometimes it acts like a particle. The same is true of the electron. (Actually, the same is true of a cannonball as it, too, has a wavelength, albeit a very short one. All masses have associated wavelengths). But, occasionally you can read, even in textbooks, arguments about whether a photon *is* a particle or *is* a wave. Such arguments lead to the *amazing duality of light*. "Is it a particle? Is it a wave? Oh my gosh, it's both—the *amazing duality of light*. Yes, well...imagine that we are trying to understand what an orange is and have little grasp of it because it is outside of our daily experience. We decide that it must be a carrot, because it is orange in color. But no, it's an apple, because it is round. Oh my gosh, it's *both* an apple *and* a carrot—the *amazing duality of the orange!* Well, no it isn't—it's an *orange*. The same can be said about photons and electrons. They are what they are. They only seem to be two things at once because in our early understanding of physics, waves and particles were thought of as two different things. We now understand them as manifestations of the same "thing."

In fact, everything might just be one "thing." Physicists are hopeful that someday all physics will be shown to derive from a single equation—a *grand unified theory* (of everything), or GUT.

Sometimes, when we come to realize that what we thought were two distinct entities are, in reality, the same thing, it can so alter the way that we look at these former "distinct" entities that we completely rearrange our thinking about them. This is sometimes referred to as a paradigm shift. We are probably going to need one of those before we can solve the supervenience paradox. But before we try to obtain that, let's first get familiar with what a paradigm shift in thinking is like by examining one brought on by a particular aspect of Einstein's theory of general relativity.

In his theory, Einstein argued that what everyone had thought to be two distinct entities (acceleration and gravity) were, in fact, the exact same thing.* At first glance, it might seem that it would be easy to tell the difference between the force you feel when the Earth pulls down on you (gravity), and the force you feel pressing you toward the floor of a rapidly rising elevator (acceleration). One seems to be caused by your motion, while the other is not.**

It would seem that there should be an easy way to show that gravity and acceleration are different. Let's try a nice thought experiment that will make clear sense of all of this. Imagine that you are in deep intergalactic space, very far from any star, and are standing in a glass elevator—not a high probability event, I grant you, but thought experiments tend to be like this. Suddenly you begin to feel a downward force. Has the gravity in the elevator increased owing to the mass of some unseen object that has just appeared below your feet (a mini black hole, perhaps), or is the elevator rapidly accelerating "upward?" Good question. How could you know? Well, imagine that I have set a beam of light a few meters away, perhaps one of those laser pointers, and I aim it so that it will strike the side wall of your glass elevator equidistant between the floor and ceiling and perpendicular to what would be any axis of hypothesized "upward" acceleration. The photons of light would shine right through your glass elevator and

---

* Technically, Einstein took issue with Newton's understanding that there were two distinct forms of mass, gravitational mass and inertial mass. Einstein's position was that there was only one form of mass; that gravitational mass and inertial mass were the same.

** Here's where it gets tricky, because Einstein showed that all motion is relative. So, if acceleration happens when you're moving, well, how do you know for sure that it is you who is moving? Or in Einstein's words, "When does Princeton arrive at the train?"

exit the opposite side equidistant from the floor and ceiling (so long as you didn't stand in the way and block the beam of light) (see Figure 14).

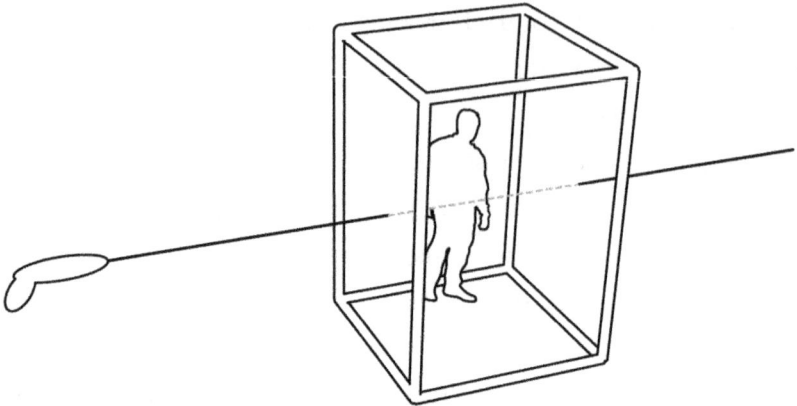

Figure 14. A beam of laser light is aimed through a stationary glass elevator that is in deep space. An observer within the elevator is able to witness it. The beam of light passes through the elevator in a straight line.

It should be pointed out that modern physics theory is strongly supported by vast amounts of solid evidence gained from decades of experiments with particle accelerators. We are not just talking about some guess somebody has pulled out of a hat. In fact, the standard model has been found to be correct in its predictions, when its predictions can be tested, down to an error of one in 10 billion.

Now picture the following. If you were accelerating "upward," when I shined my laser beam, your elevator would be struck by the entering photons of light equidistant between the floor and ceiling as previously stated, *but* while the photons of light were crossing your elevator toward the wall through which they will exit, your elevator will have *moved* upward. This causes the traveling photons (which go in a straight line) to exit your elevator closer to the floor than the ceiling (see Figure 15.1). In fact, from your perspective inside the elevator, the beam of light will appear to bend, or curve downward like a bow, owing to your upward acceleration (see Figure 15.2).

Now, let's look at the same situation if we apply a gravitational field to the area below your elevator so that it yields the same "downward" pressure on you as did the acceleration we just tried. This time, however, your elevator doesn't move even though you feel your body being pulled downward. Once again we shoot the laser beam through your elevator. But this time the laser beam does not bend, why, because photons have a mass of zero and are, therefore, not attracted by a gravitational field. Well, there you go! Our thought experiment has shown that there is a way to tell the difference between gravity and acceleration after all, and that they are not the same thing. In a gravitational field, the light will not appear to bow, or bend, from the vantage point of the observer, but during an acceleration it will.

But Einstein could do the math, and his relativity equations were telling him that acceleration and gravity were the same. But how can we reconcile this with the results of the thought problem? Einstein had an interesting answer. He argued that the gravitational field would in fact bend the beam of light! This would yield the same result as that obtained by acceleration in the glass elevator. But how could that be when it is well known that photons of light have no mass and are not attracted by a gravitational field?

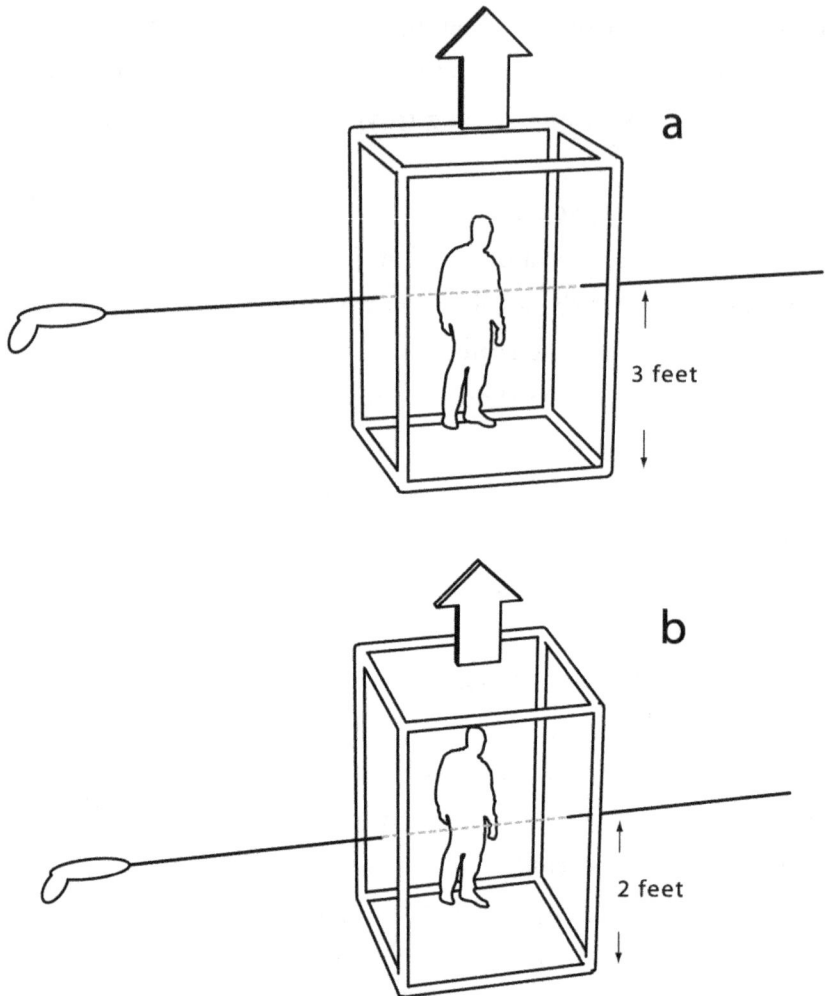

Figure 15.1. A glass elevator with an observer travels "upward" while a laser-pointer beam is aimed through the elevator. In (a) we see that the beam of light passes straight through while the elevator moves upward. In (b) we see that from an outside observer's point of view that the elevator appears to have moved upward through a steady beam, so that the beam is now closer to the floor of the elevator.

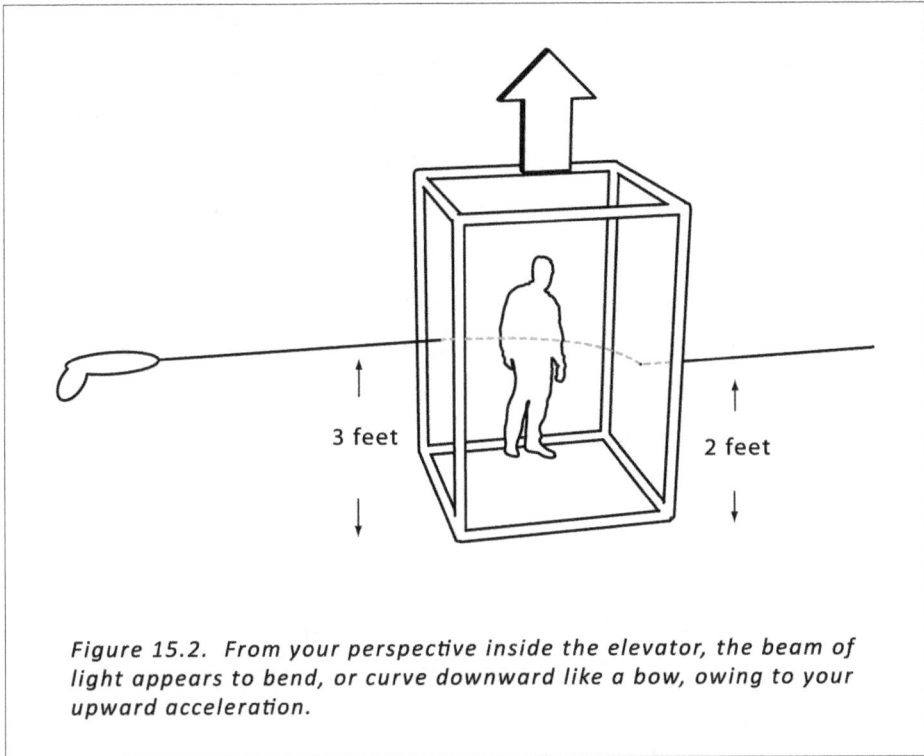

Figure 15.2. *From your perspective inside the elevator, the beam of light appears to bend, or curve downward like a bow, owing to your upward acceleration.*

There could only be one answer. If light were not attracted by gravity, and therefore would be expected to travel in a straight line through the gravitational field, the only way one could explain its bending would be to assume that the *space* the light traveled through was being bent! In that way, the light would bend even though it was going "straight," because the space it passed through was being curved.

Intuitively, this argument makes no sense. How could space be bent? But as you recall, space is a thing, it is not nothing (recall that if there were *nothing* between the Earth and the Moon, they'd be touching). So, if this "thing" we call space could somehow be bent by gravity, was there any way to prove it?

As it turned out, there was a way. On May 29th, 1919, a total eclipse of the sun was to occur. At the moment of totality, the sun would be in front of a fairly bright field of stars (part of the Hyades group). If the distance between two bright stars could be determined beforehand, two stars that would be on either side of the sun during totality, that particular measurement could be compared with the same measurement taken when the sun was between the two stars. If a large gravitational field truly "bent" space, then one could predict how much deflection one might expect the stars to have from their normal positions when their light was passing by the darkened sun (see Figure 16).

To get the best possible measurements, Sir Arthur Eddington, one of the most prominent astrophysicists of his time, arranged for British Expeditions to Sobral, Brazil and to Principe in West Africa. In Eddington's own words, "The results from this [photographic] plate gave a definite displacement, in good accordance with Einstein's theory and disagreeing with the Newtonian prediction." The shock of this discovery can be sensed from the headline in the New York Times following Eddington's report:

LIGHTS ALL ASKEW
IN THE HEAVENS

MEN OF SCIENCE MORE OR LESS AGOG
OVER RESULTS OF ECLIPSE OBSERVATIONS

Einstein Theory Triumphs

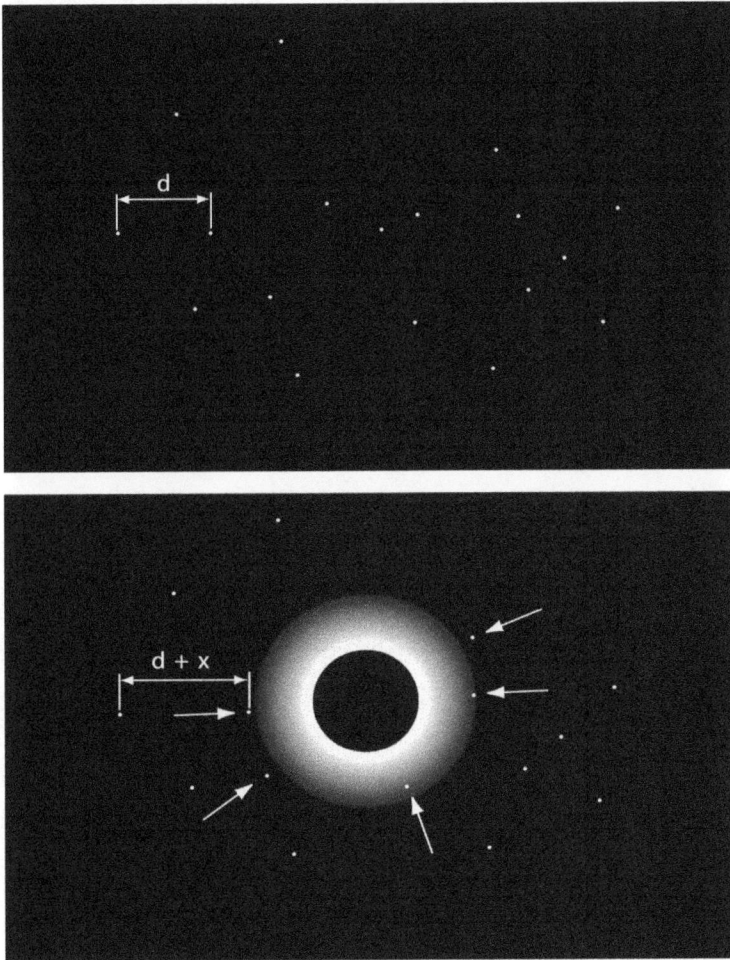

Figure 16. In the top image, we see an arrangement of stars in the night sky as might be observed by someone on Earth. The distance between two of the stars is given as "d." In the bottom image, we see the same stars during a total eclipse of the sun. Because the sun is now between the stars and the observer, we can use Einstein's equations to predict the deflection of the starlight owing to the power of the sun's gravity to warp the space through which the starlight passes. Stars closest to the sun look as though they have changed position most, appearing to have moved toward the sun (the direction of apparent movement is indicated by arrows). The distance previously measured as "d," is now observed to be "d+x."

Light, although massless, *did* bend in a gravitational field. Einstein's equations had shown that space could be bent and twisted, and even that time could flow at different rates. Three-dimensional space had now become four-dimensional space/time, and this led to a paradigm shift in thinking if ever there was one. From that day forward, astronomers had a different and fundamentally new understanding of the universe.

A similar example can be seen from the work of the great modern physicist Stephen Hawking. A black hole, the possibility of which was first described by J. Robert Oppenheimer, is a collection of mass so dense and substantial that its gravitational field prevents anything from escaping from it, even light (thereby causing the object to appear as a black spot or hole in space). If we think of a black hole in terms of Einstein's theory of general relativity, we would say that the space surrounding a black hole was so warped that light leaving the black hole (and continuing to follow a "straight" line) would follow the curved space right back into the black hole. Nothing leaves a black hole, ever; or so one would assume.

However, if you look at black holes in a different manner, it turns out that they can radiate away their mass, bit by bit, until they actually "evaporate" to nothing. But since nothing can ever escape from a black hole, a radiating black hole would appear to be impossible. I mean, what could it possibly be radiating out into space that wouldn't just get bent right back into it? That seems like a reasonable question, all things considered.

The *Heisenberg Uncertainty Principle* tells us that the energy state of even absolutely empty space cannot be zero. There is no uncertainty about zero—it's zero. So, the Uncertainty Principle shows us that there is always some energy fluctuation, even in a vacuum, along a probability gradient of possible energies that approaches zero. So, this leaves us asking, how does the energy of "empty" space manifest itself? Where does the energy come from? How can there be any energy in totally empty space? The problem can be solved if we hypothesize the existence of very tiny particles (and their concordant anti-particles) that rapidly appear and then annihilate each other (since particles and anti-particles will do that to one another when interacting). This process is then repeated endlessly everywhere and at all times. The idea, then, is that these virtual particles (as that is what they are called) are everywhere, forever appearing, interacting, annihilating each other, and then reappearing. You can think of them as the fabric of the thing we have been calling "space." Because these virtual particles are doing this all the time and everywhere, even the energy level of "empty" space will fluctuate a tiny bit, enough to be as unpredictable as Heisenberg's equations would say that would be. In fact, as you recall, it is from such vacuum fluctuations that even a "big bang" can be generated.

This might all seem a bit fanciful and convenient, if it weren't for the fact that the existence of virtual particles has been confirmed by experimental evidence. If virtual particles did exist, physics says that two uncharged metal plates brought very close to one another (within a couple of atomic diameters) should be "attracted" to each other even though no magnetic or electrical forces were directly involved, simply because the density of the virtual particle interaction will be less between the plates than it is in the area surrounding them. If there were no interactions between virtual particles then there should be no differences in the densities of the space surrounding the metal plates. Experiments have shown that there is a density differential, and that the metal plates, when brought close enough together, will be drawn toward one another. This has come to be known as the *Casimir Effect*.

In the early 1970s, Stephen Hawking wondered about the *event horizon* of a black hole and what such a horizon might do to this virtual particle fabric. The event horizon of a black hole is the exact point beyond which anything closer to the black hole has no hope of escape. In other words, if you pilot your ship across an event horizon, well, you're going in and that's that. Hawking calculated that the event horizon would be so powerful that right at its edge it could tear apart the coupling of virtual particles, drawing one of the particles in and leaving the other half of the pair stranded (see Figure 17).

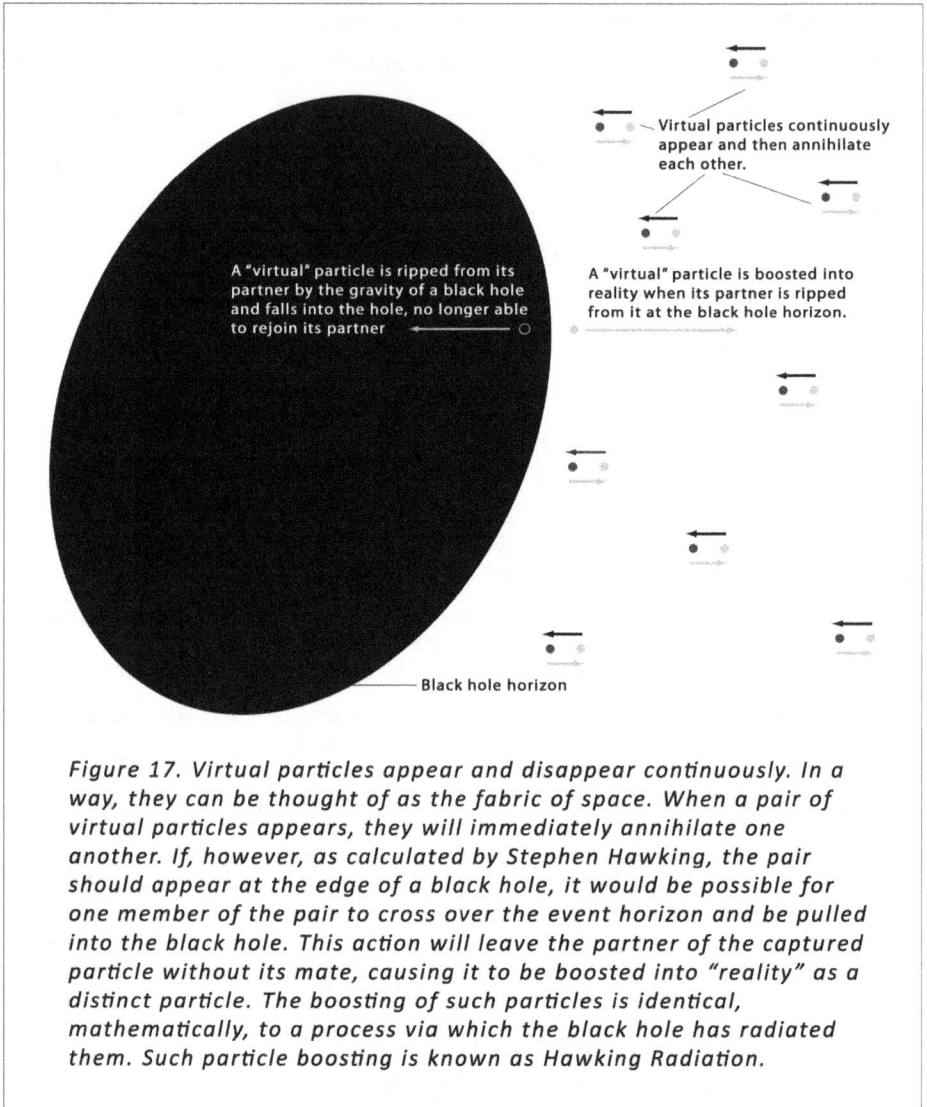

Figure 17. Virtual particles appear and disappear continuously. In a way, they can be thought of as the fabric of space. When a pair of virtual particles appears, they will immediately annihilate one another. If, however, as calculated by Stephen Hawking, the pair should appear at the edge of a black hole, it would be possible for one member of the pair to cross over the event horizon and be pulled into the black hole. This action will leave the partner of the captured particle without its mate, causing it to be boosted into "reality" as a distinct particle. The boosting of such particles is identical, mathematically, to a process via which the black hole has radiated them. Such particle boosting is known as Hawking Radiation.

When this occurs the stranded particle is "boosted" into becoming a real particle. Its partner in the dance of annihilation is gone, so it continues to live and perhaps even escape from the black hole.[*]

At this point, Hawking applied the principle of equivalence as he realized that there was no way to tell, in terms of physics, any difference between escaping boosted particles and a radiating black hole. In fact, the force used by the black hole to boost the particle saps it of energy or mass exactly as though it had radiated the particle. For those who are curious, the power emitted by a black hole in the form of Hawking radiation (as it has come to be called) is given by the formula $P = \hbar c^6 / 15360\pi G^2 M^2$, where P is the power of the radiation, $\hbar$ is the reduced Planck's constant in Joules per second, c is the speed of light, G is the gravitational constant, and M is the mass of the black hole. Not to worry if this particular formula makes little sense to you, the important thing is that whether you are calculating the "evaporation" of a black hole owing to radiation, or the energy it uses boosting virtual particles, you would use the same formula! In fact, if the black hole is small enough, it might eventually evaporate below black hole status and flash back into bright existence with a bit of a bang (not a huge bang, as bangs in the universe go, but big enough to possibly be observed by astronomers).

---

[*] The hypothetical "froth" that we discussed earlier that was caused by the expansion of space, may well be due to the boosting of virtual particles. If virtual particles appear, they might have difficulty annihilating each other if during their brief appearance the space between them expanded rapidly. Such rapid expansion might suddenly leave both the virtual particle and its complementary virtual anti-particle suddenly too far apart to rejoin, thereby boosting them into "reality." For reasons not well understood, it appears that anti-matter is not the mirror image of matter and that anti-matter tends to break down over time leaving our universe composed mostly of matter. Much of this matter might be from virtual particles boosted into existence by an extremely rapid expansion of space.

We now see that the seemingly impossible can become possible, if we can find an equivalent way to look at phenomena and thereby achieve a paradigm shift in our understanding of it. Light does bend in a gravitational field and black holes might radiate, even though neither seemed possible at first. Could the principle of equivalence be applied to our understanding of consciousness so as to eliminate the supervenience paradox?

THE OBVERSE WORLD

All is riddle, and the key to a riddle is another riddle.
— Ralph Waldo Emerson

Coin collectors often refer to the most prominent side of a coin as the obverse side. The other side is often referred to as the reverse side. Implicit in the reference to an obverse face is the understanding that there is another side to the *same* coin.

I want to introduce you to two worlds. Let us call the first one the obverse world. In the obverse world you are born, you live a life, and eventually you die. In the obverse world, you are always you. You never discover that somehow you have become someone else. You are born you, you stay you, and you die as you. Your consciousness resides for a complete lifetime within one body and that body is always yours.

If the obverse world sounds familiar, it should, because I am describing the world in which we all live. There is little reason, therefore, to continue with a more detailed description of the obverse world, because you are already fully familiar with it.

But now I want to tell you about another world. It is a very strange world with which you are not at all familiar. It is a world in which strange things happen and consciousness is unlike any phenomenon you have ever imagined.

This other world in all respects looks like the Earth. It has the same people and places as the Earth. The difference between the two worlds lies in the way that consciousness manifests itself. On this other world, although there are 6 billion people, there is only *one* consciousness; all the 6 billion inhabitants share that one consciousness. In essence, all 6 billion people are the same "person."

At this time I don't wish to describe in detail exactly how such a consciousness might operate, but would rather, just for now, use a simplified model to outline its properties so that I can easily make the point that I wish to discuss. So bear with me while I take brief liberties with physics in order to outline this concept.

Imagine that this one consciousness (the sense of self-awareness) "jumps" from person to person, stopping briefly in each person before moving on. Let's say, for the sake of argument that it will reside in each person's brain for one full second before moving on and "jumping" to the next individual, where it will reside for another full second, and so on. (I realize that none of this makes any rational sense. I am only attempting to introduce this rough concept first before attempting to place it in a realistic and believable context). At any given time in this world, only one person will be conscious, and that would be the person in which the one and only consciousness is residing at that particular second. The other people will continue to function as zombies until they get to have their one second of consciousness. My question to you is, "What would it be like to be the consciousness in such a world?"

At first it might seem that your life would be impossible to lead because you would be someone different every second of every day. Imagine it. One second you are a little girl on a bike, the next second you are middle-aged baker kneading dough, the next second you are a teenage boy reading a book, and the next second you are an elderly woman playing bridge with friends, and the next second you are a toddler taking a nap, and the next second—well, you get the picture. It would be an impossible blur of a life that would make little sense.

However, I will argue in a thought problem that we can apply the principle of equivalence in this instance. Allow me to remind you again at this point not to concern yourself with the idea of a consciousness that "jumps" from brain to brain, or that lasts in each targeted brain for a full second. The concept of a "jumping" consciousness, or the idea of it lasting in any one person for a full second, is only a contrivance for purpose of example. I don't actually expect a consciousness to really "jump," but we will deal with such particulars later. For right now, though, I wish to argue that if we lived in such a world, it would be, in fact, indistinguishable from our own "obverse" world. It would simply be another side of the same coin. If you lived in the world of the "jumping" consciousness, the world in which we are all the same personal consciousness, you would be unable to tell it from the world in which we currently reside no matter what you did.

At first this seems preposterous. But let us begin by chipping away at our grip on the obverse world. We're not going to throw away the coin, only turn it over. It is still that same coin. To begin with, I am quite sure that I have always been me and that I have never been someone else. I am sure that you feel the same way about yourself. You have always been you and have never been someone else. But how do you know? Einstein once looked at the face of a clock on a clock tower, and thought about the light that reflected the image that showed the time of day. He wondered what it would be like if he could jump aboard the photons of that image and ride along with them at light speed as they raced away from the clock. He realized that as long as he rode along with that image that every time he looked at the image it would stay the same time. Time would have been frozen at the moment he leapt aboard. If he slowed down a little, the later images of the clock face would begin to overtake him and the hands of the clock would slowly begin to move. Time would begin to creep forward as he slowed. In this imaginary example we can see the relationship between velocity and time. The greater our velocity, the slower time moves. Experiments have proven this to be true in exactly the way Einstein predicted.

So let us imagine that we are the consciousness of this "reverse" world. You (the consciousness) leap into the brain of someone. But this single consciousness has no memories. The consciousness is only the generation of self-awareness, the sense of being alive. If the consciousness is aware of who it is, it is only because it has tapped the memory that resides in the brain it has just sampled.

But, you argue, you are you, and you have always been you. But again, how do you know? I understand that you are conscious now, and seem to be "you." But how do you know that you weren't someone else a few seconds ago? Perhaps you remember doing something yesterday. But did your *consciousness* actually experience that? Were you this particular person yesterday? All you can say is that by being conscious now you can tap into the memory of the brain you are in and bring up a recollection. You might only have been this person for the last second or so and will be someone else a second from now. If such a thing were happening, you would never be aware of it. You would always feel at home in the body you were currently in because you would tap into that brain for information. In a sense, we have touched on this idea before when we considered that consciousness might come and go and that during the times it was "gone" your body might continue to function normally in every way, only as a so-called zombie. But, no matter what experiment you might try to imagine, there seems to be no way to ever know that you have always been you and not been other people as well. You can never know anything for sure other than you are you right at this very instant. You could very easily have been someone else a second ago and be yet another person a second from now and never be aware of it in the slightest. This strange reverse world and the obverse world in which we live are equivalent. There is the temptation to ask, "But which is the real world, the obverse world I know well, or this strange reverse world?" But we are just looking at the same experience from a different angle. Both are just different ways of looking at the same phenomenon, different ways of seeing exactly the same "thing." If there is no way to tell one world from the other, and there isn't, they are obviously the same world. At this point in our journey, though, I would imagine that you would prefer to accept the obverse world as the better view because there is no evidence that a "jumping" consciousness could exist according the laws of physics. Furthermore, you may have noticed that this reverse world we have been considering in no way overcomes the supervenience paradox.

However, this reverse world idea that we are toying with, actually can lead us to a solution of the supervenience paradox and, in turn, a number of other stunning possibilities.

ONE CONSCIOUSNESS FITS ALL

We all know that something is eternal.
And it ain't houses, and it ain't names, and it ain't earth,
And it ain't even the stars...
Everybody knows in their bones that something is eternal,
and that something has to do with human beings.
　　　　　　　　　　—Thornton Wilder, "Our Town"

I have been arguing all along that consciousness is of the body, and it might well seem that this central precept has now been derailed by a discussion of a "jumping consciousness." After all, when we start to talk about a consciousness that "jumps" about, it certainly sounds like I am conjuring up some sort of spirit that floats from body to body. However, I only used that image so as to easily explain the concept that you might not always be you, and that you might be other people as well, and that you would never be aware that such a thing was occurring. What we need then, is some indication from science that there might be a way for shared consciousness to exist without evoking floating spirits before we can feel comfortable considering this concept, and then thinking of a possible way to implement it to avoid the supervenience paradox.

Einstein was fortunate that there was a way to test his theory. That is why relativity is considered to be physics rather than metaphysics. Einstein's theory was part of a testable science. Hawking's theory that black holes radiate is also testable, although no test has yet been conducted. Such a test would require some proximity to a black hole and a way to accurately measure any radiation being emitted from it. The fact that it can be tested, even if not right away, also places this theory within the purview of hard science. Interestingly, sometimes physicists wander off some considerable theoretical distance in mathematical pursuits of answers and occasionally end up with theories that are not testable in any obvious or immediate way. Modern string theory is one such example. String theory is an effort to combine the molar with the molecular. Our understanding of gravity helps to explain things on a large or molar scale, while our understanding of quantum mechanics works well for explaining things on the very small atomic and subatomic scales. Unfortunately, gravitational theory and quantum mechanics do not appear to mix well. What works well in the big world appears to be quite foreign in the world of the tiny and vice-versa. String theory, which describes all particles (electrons, muons, protons, etc) as manifestations of differing "vibrations" of super tiny "strings" (roughly a millionth of a billionth of a billionth of a billionth of a centimeter in length), shows great promise for unifying gravitational and quantum theories. However, many physicists view string theory as a metaphysical pursuit since no clear ways to test string theory have been developed. Physicists such as Brian Greene and others are, however, searching for a way to test the basic assumptions of string theory so that they might bring it into the realm of science.

Consciousness, especially a shared consciousness, grimly defies any form of empirical testing. Part of the problem lies in the fact that one must be conscious in order to observe empirical events. Consciousness, therefore, can't be isolated and examined without interfering with empirical observation. Curiously, because of this fact, one might correctly argue that all empirical observation, and by extension all science, is faith-based, inasmuch as one has to have faith that what he is empirically observing is real. Recall Descartes dictum, "I think, therefore I am." Therefore, even if I had the advanced neurological tools to penetrate the deepest recesses of the brain, I doubt that consciousness would ever yield fully to the requirements of science as an area of proper study and will most likely always have at least one foot firmly planted in the metaphysical. This doesn't thwart our pursuit, however, because as you recall I said right at the start that I believe what happens to us after death is unknowable. All we can hope to do is make the best guess, and perhaps argue that the common assumption that "when you're dead, you're dead and that's it" actually makes little rational sense. That alone, I think, is pretty interesting.

To return to our discussion: Imagine then, a flip side to our obverse world, a reverse world in which we are all the same person (by "reverse" I in no way mean to imply that things run backwards, but only use that term as coin collectors do, to denote the backside of a coin). To accomplish this in a logical and rational way, we will need to conceive of how all of us could reasonably share one consciousness. To do so, we will need to turn once again to modern physics for an indication that a shared consciousness might somehow be a manifestation of the workings of space and time.

SPOOKY ACTION AT A DISTANCE

I've made up stuff that's turned out to be real, that's the spooky part.
—*Tom Clancy*

If I drop a pebble into a pond it is not surprising that it should cause ripples to form in the water. These ripples might radiate outward and eventually make a leaf floating on the surface of the pond move up and down as the ripples pass it by. In physics this is known as a local effect. The ripple might even cause the water to move in such a way so as to reflect a photon of sunlight off in the direction of Pluto, a photon that might reach that distant planet some hours later after its light-speed journey. This, too, if it should occur, would all be considered a local effect—all part of the understood mechanics of physics.

A non-local effect might occur as follows. You drop a pebble into a pond and it causes *at that exact same moment* a little chunk of ice on the surface of Pluto to bounce. Of course, the world doesn't work that way. Physics isn't non-local. If it worked that way it would make no sense at all. Non-local effects would be just plain spooky. And that is exactly how they were perceived—when they were discovered.

In 1935, Einstein, Podolsky, and Rosen published a paper in which they pointed out that quantum mechanics (an area of physics with which Einstein often took issue) would, under certain circumstances, predict the occurrence of non-local events. Einstein referred to this as, "Spooky action at a distance." With the publication of their paper, Einstein and his colleagues were not predicting that non-locality would be discovered, but rather that the fact that such a deduction could be derived from the mathematics of quantum mechanics was an indication that quantum mechanics was a flawed system. Einstein never did accept the basic precept of quantum mechanics that the world was indeterminate. Einstein rejected

Heisenberg's Uncertainty Principle as a fundamental law of nature with the famous comment, "God does not play dice with the universe." Niels Bohr, the physicist who described quantum mechanics and who once rejected a particular quantum theory on the grounds that, "It wasn't weird enough to be right," responded to Einstein's remark by saying, "Don't tell God what to do."

In 1964, John Stewart Bell published a theorem in which he suggested a way to test non-locality and predicted that it would occur. It took five years before the test could be set up and conducted. The "spooky action at a distance" that Einstein had derided as a flaw in quantum theory was, in fact, a description of how the real world worked! Professor Henry Stapp, a physicist at the Lawrence Berkeley National Laboratory, has called Bell's Theorem "the most profound discovery of science." You will note that the words he chose were "the most profound discovery of *science*," not just physics, and many scientists agree with that assessment. It is amazing to me that a typical student might complete high school, or even college, and never hear of Bell's Theorem. The most profound discovery of science and hardly any laymen seem to know about it.

In 1969, Clauser, Horne, Shimony and Holt, proposed an experiment that confirmed Bell's Theorem. Their experiment has been repeated many times and in various forms. The theorem is always supported. What they found was so spooky, so bizarre in fact, that it has left physicists in a permanent state of jaw-drop.

For the next few pages let's look at the experiment. The set-up and process is fairly simple and straightforward. The results are, well.....the results are another thing altogether. The original experiment was conducted with photons, but other experiments have since been run using electrons, protons, and even charged atomic nuclei. The same spooky outcome is always the result.

To begin, imagine a simple arrangement in which a beam of photons is fired into a splitter, which then separates the beams into two beams of light. A splitter need not be anything fancy. A partially coated mirror will do. Such a mirror would allow about half of the photons to pass through while reflecting the other half along a second path. Using sets of fully coated mirrors, the beams of photons can then be rejoined and measured at a final detector (see Figure 18). All in all, the photons will have traveled a few meters. There is nothing sophisticated about such an arrangement and any capable high school student should be able to set up such an apparatus. Things get a little more interesting when a *single* photon is fired through the apparatus. Firing just one photon requires a little more sophistication.

Interestingly, if a single photon is sent through the apparatus, it will follow *both* paths shown in Figure 18, showing nicely that sometimes a photon will act like a wave. In fact, when many photons are used, and the paths are combined at the final destination, it is observed that the photons can interfere with each other causing the standard wave interference pattern to emerge (see Figure 19).

A single photon can, amazingly, interfere with itself! But this is to be expected in the quantum world. Photons act like both waves and particles; they are exceptional unto themselves and uniquely what they are. We have been down this road already and will simply allow that one photon can travel two paths at once by acting like a wave that fans out over both pathways and then interferes with itself when the paths are recombined.

But let's start to make our experiment really interesting. Photons, as waves, come in types, depending on the orientation, or polarization of the wave. You can sort them from each other. In fact, polarized sunglasses do just that, allowing in only waves that are oriented at a certain angle. You might think of them as the waves that are moving side to side, rather than the ones that are moving up and down (see Figure 20).

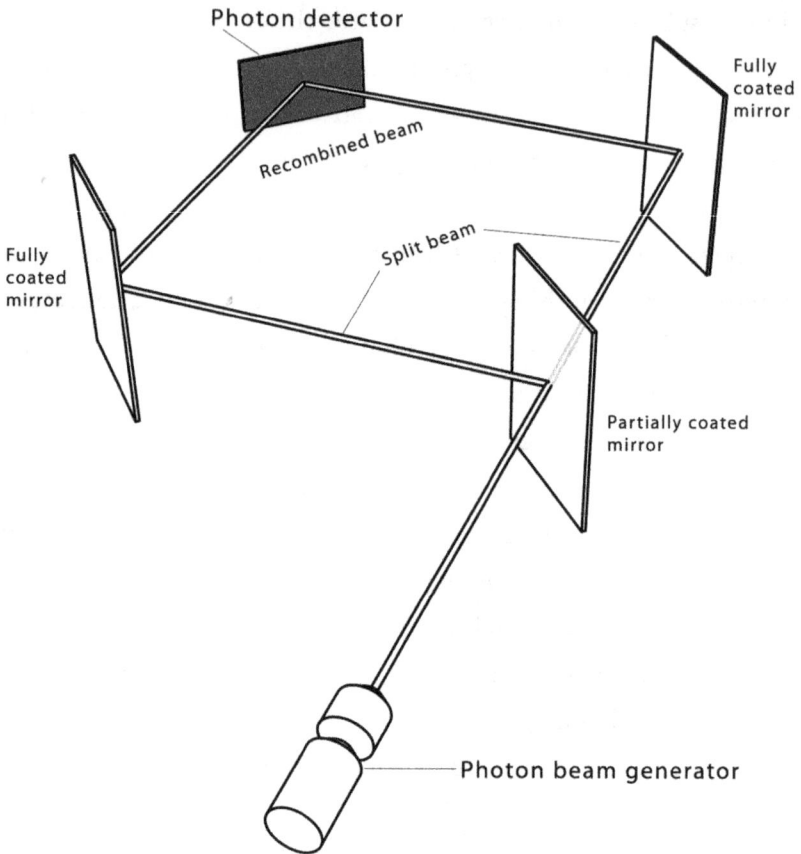

Figure 18. A beam of photons is sent through a splitter (a partially coated mirror), which then separates the beams into two beams of light. The splitter causes about half of the photons to pass through while reflecting the other half along the second path. Using sets of fully coated mirrors, the beams of photons can then be rejoined and measured at a final detector.

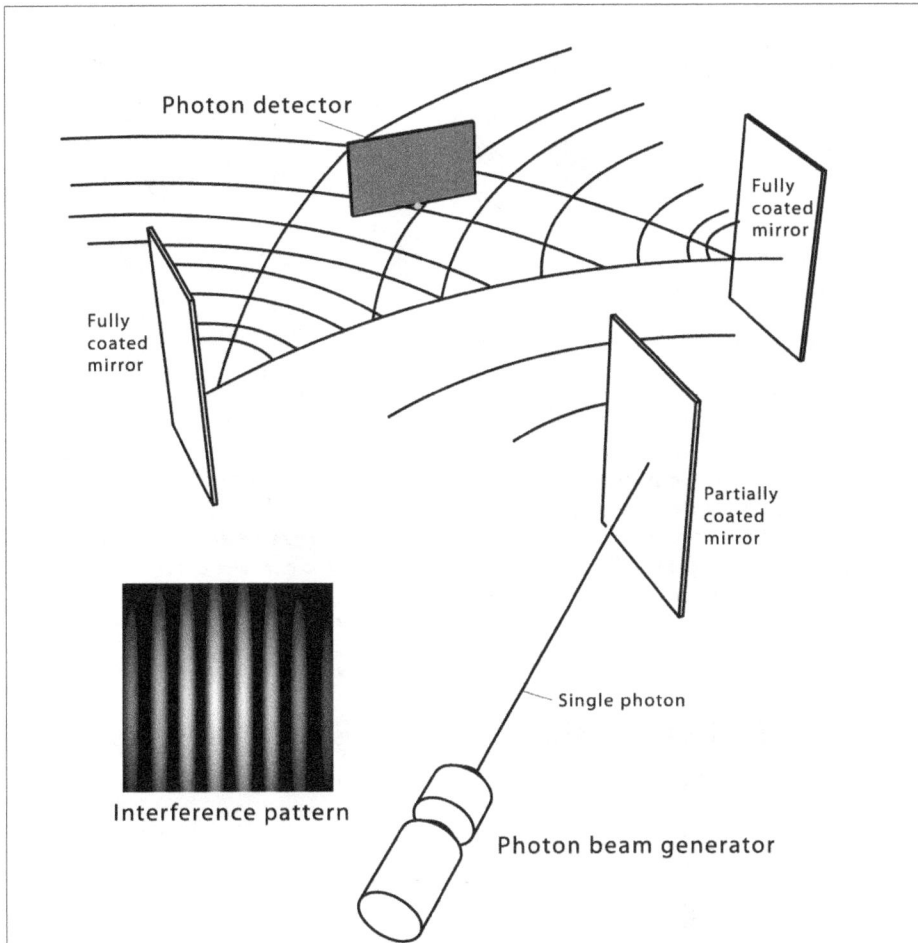

Figure 19. A single photon can follow both paths, showing the wavelike properties of the photon. The single photon "travels" to both mirrors, and can be represented by an ever-expanding wave front. When the wave reflects off of both mirrors, it interferes with itself before reaching the detector, at which point the wave collapses and the photon hits the detector as though it were a particle (gray spot). Evidence that the single photon has traveled both routes comes from the fact that the photon can be observed to interfere with itself by causing a classic interference pattern of light and dark stripes (see photo insert).

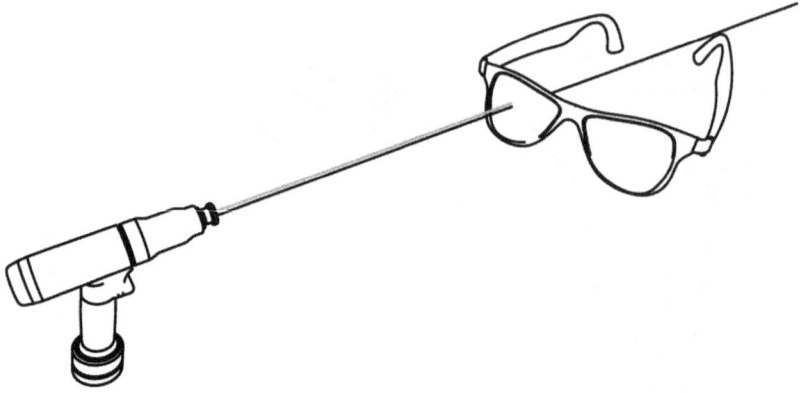

Figure 20. A simplified example of a polarized filter: Light waves oriented at different angles (represented by the two colors of beams fired from our ray gun) approach a pair of polarized sunglasses. The glasses, acting as a filter, only allow waves of a certain orientation to pass, thereby cutting down on glare.

The only point that I wish to make here is that photons come in different varieties. The same is true of electrons; they have an intrinsic angular momentum that is referred to as spin. If you wish, you can separate electrons that have a +1 spin from those that have a –1 spin and those that have a zero spin.

It is, thankfully, not important for our purposes to get into photon polarizations or electron spins. What is important is for you to know that photons or electrons come in different varieties and that they can be separated from each other. For the sake of simplicity, and so that we can better concentrate on the effect of the experiment rather than its intricacies and particulars, let's just discuss photons and say that they may be sorted along two dimensions; black vs. white, and rough vs. smooth. (I will use these terms in place of different forms of polarization and other physical attributes). So, in our example a photon may be "black" or "white," "rough" or "smooth." As you can see, this leaves us with four possible combinations of photons; those that are black and rough, black and smooth, white and rough, or white and smooth.

What's most interesting at this point is that Heisenberg's Uncertainty Principle allows us to know whether a particular photon is black or white, or if it is rough or smooth, *but not both at the same time*. Once I find out it is black, that very measurement precludes me from finding out its roughness. Or, conversely, if I measure that it is either rough or smooth, I will be forbidden from knowing if it is black or white. (As you will recall, Werner Heisenberg's Principle of Uncertainty clearly shows that it is not possible to ever know both a particle's exact location and its exact momentum at the same instant. Our example of rough/smooth, black/white represents various similar parameters as they pertain to spin, polarity, and other phenomena that are also mutually excluded by the uncertainty principle.)

To begin our experiment, we will place a device in the path of the stream of photons (the stream of photons is a simple beam of light). This device can weed out the "black" photons from the "white" ones. (Although you might be imagining some complex apparatus, recall that a simple pair of polarized sunglasses can accomplish this feat). The beam that is allowed to pass reaches a detector, and the detector records that there are now only "white" photons, because the "black" ones were filtered out (so far, so good). None of this should come as a surprise. It is what we would expect to see happen. In fact, you can do the same thing and use a filter to separate "rough" photons from "smooth" ones. This additional fact, however, leads to a very interesting idea that should allow us to overcome Heisenberg's Uncertainty Principle!

Let's set up this next experiment and forever be rid of that nasty uncertainty principle. Here's what we do. We fire a beam of photons through a filter that separates out black photons from white ones. Whether they are also rough or smooth doesn't matter for now. Once we have separated out the black ones, the white ones then travel on as we have seen happen before. Next, and this is the ingenious part, we pass the "white only" photons through a filter that removes the "rough" photons from "smooth" ones! So now we will have only white *and* smooth photons, which is not allowed, since the uncertainty principle forbids us from ever knowing both of those qualities at the same time for a given photon.

So we run the experiment. The beam passes through the first filter. We measure that the black photons are reflected away while the white ones pass through. Then the white photons pass through the second filter. We measure that the rough photons are reflected away allowing only the smooth ones to pass (we've got em' now!). Finally the *white and smooth* photons reach the final detector, which reports that while all of the photons are smooth, half of them are black and half are white!! (see Figure 21). This immediately gets filed under the heading of "What the #$%*??"

That's bad enough, but the next part of this investigation gets just plain scary. As you recall, when a single photon goes through a splitter it acts like a wave and appears to split, and when rejoined can interfere with itself. A stream of photons passing through a splitter and then rejoined also acts like waves. The waves seem to expand outward at the speed of light across both split beams and then "rejoin" when the wave front reaches a target (in this case our detector). At that point, the photons act like particles and yield all their energy to that one location. To imagine something like this in the molar world with which we are all familiar, picture dropping an empty soda can off a pier in New York into the Atlantic Ocean. A ripple caused by the falling can then spreads out into the ocean. The wave expands in an ever-widening arc. Meanwhile, there is an empty soda can floating right at the base of a pier in Lisbon, Portugal. The ripple eventually reaches that soda can. Instantly, the great arc of the ripple stretching from the equator to the arctic vanishes, and the soda can in Lisbon leaps out of the water to land on the pier.[*]

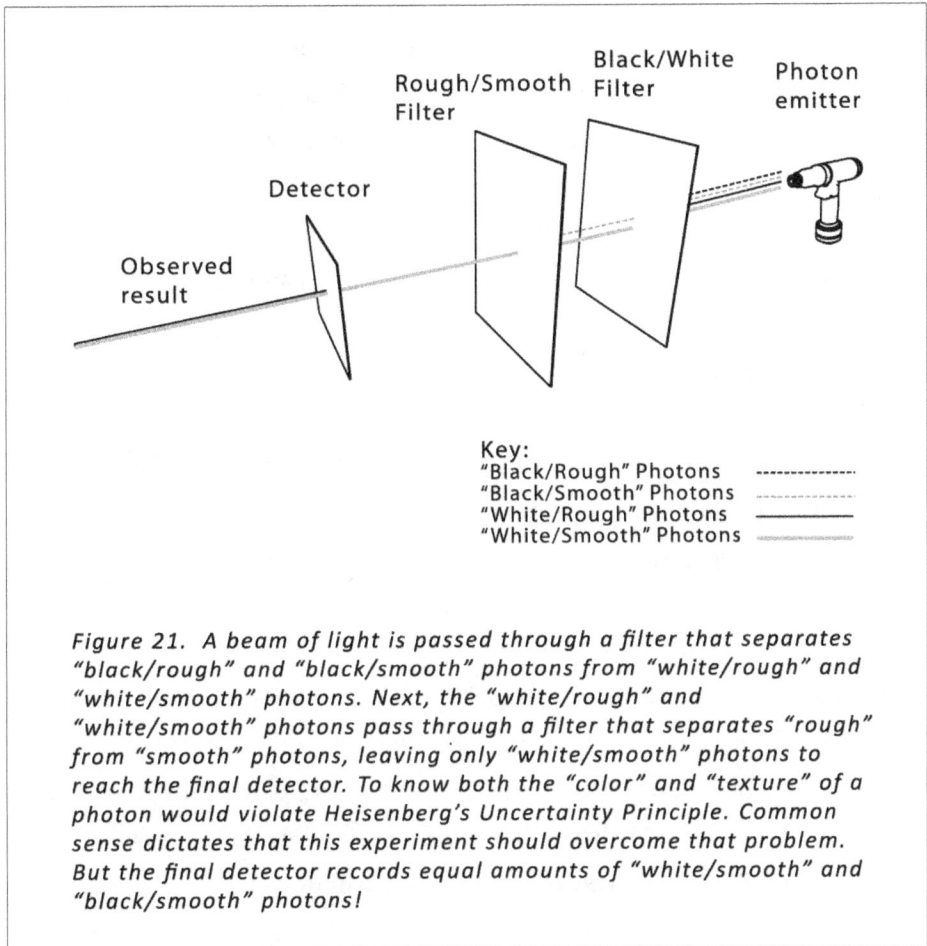

Figure 21. A beam of light is passed through a filter that separates "black/rough" and "black/smooth" photons from "white/rough" and "white/smooth" photons. Next, the "white/rough" and "white/smooth" photons pass through a filter that separates "rough" from "smooth" photons, leaving only "white/smooth" photons to reach the final detector. To know both the "color" and "texture" of a photon would violate Heisenberg's Uncertainty Principle. Common sense dictates that this experiment should overcome that problem. But the final detector records equal amounts of "white/smooth" and "black/smooth" photons!

---

[*] In reality, such a ripple would encounter interference from so many waves and currents that it would have long since vanished before ever reaching the arctic, the equator, or Portugal. This example assumes a perfectly placid body of water, and even then, there are questions about how far any ripple would extend. In the quantum world, the wave does not diminish while passing through a vacuum and imparts all of its energy to the contact target.

Who said that the quantum world wasn't strange? Keep that in mind while we set up the next experiment.

Okay, so the photons want to play games. Somehow, some way, they won't let go of the uncertainty principle even though logic dictates that they should. Even though we removed all the black photons, our final detector showed that half the photons were black anyway, and for no other reason than we removed the rough ones from a white only group. It is almost as though the photons are keeping tabs on what we are doing and going out of their way to thwart us! But that can't be. Photons can't "know" what's going on. Let's try the experiment across a wave front and see what happens.

We start with a beam of photons. We pass them through a filter that removes all the "black" ones. This leaves us with a white only beam. Now, we split that beam sending white photons off in beams that are heading in opposite directions, east and west. We know that these beams can interfere with each other when rejoined, even if we are only using one photon. So somehow these split beams are part of the same "thing." I will hypothesize that they are sharing a "wave front." The east and west beams are, therefore, like the *one* wave ripple from our earlier example that was both in the arctic and at the equator at the same time. They are part of the same ripple, or "thing." So what will happen if we do the following? Look at Figure 22, and you will see our next plan. What we will do is to sort out the beam that's heading east, but measure *the other beam*, the one that's heading west.

The "white only" eastward beam now passes through a filter that sorts out the rough photons. So now the eastern beam should be both white *and* smooth, but we know better because we've tried that before. But what we plan this time is to measure the *western* beam. It too is "white only," but hasn't passed through a rough/smooth sorting filter. According to our wave front hypothesis, the western beam might be affected by the measurement of the eastern beam anyway, since, in theory at least, the split beam is one in the same "thing," whatever that "thing" might be. And, sure enough, when we measure the western beam it shows photons that are half black and half white even though we only did a rough/smooth sort on the eastern beam.

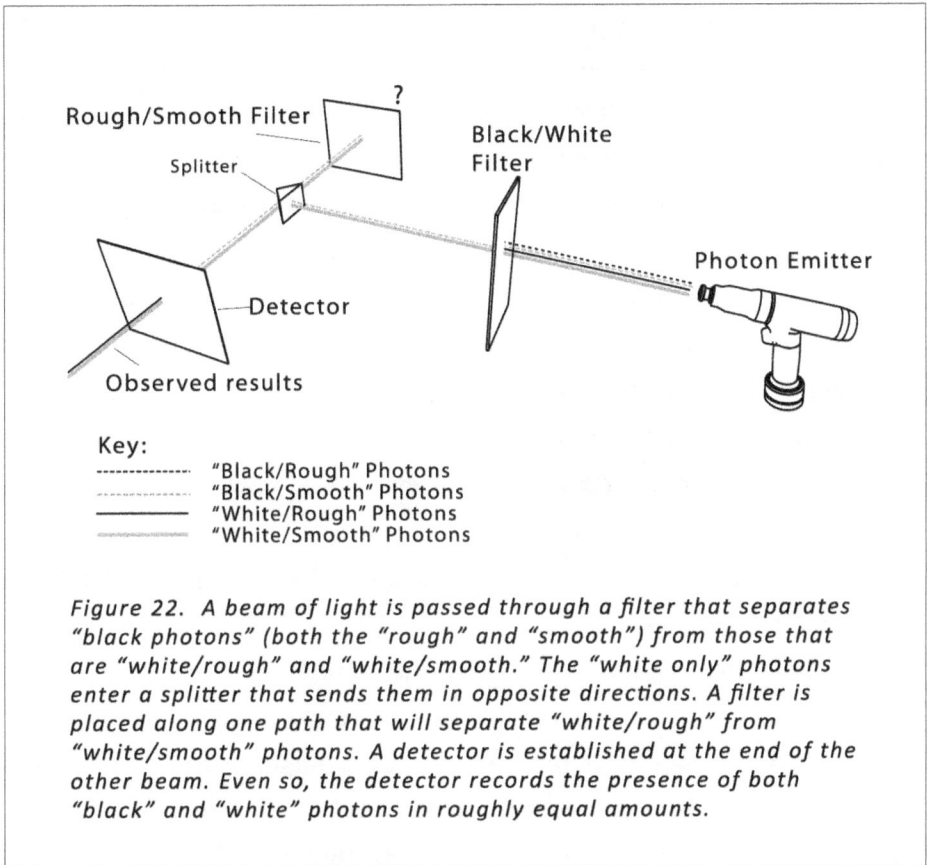

Figure 22. A beam of light is passed through a filter that separates "black photons" (both the "rough" and "smooth") from those that are "white/rough" and "white/smooth." The "white only" photons enter a splitter that sends them in opposite directions. A filter is placed along one path that will separate "white/rough" from "white/smooth" photons. A detector is established at the end of the other beam. Even so, the detector records the presence of both "black" and "white" photons in roughly equal amounts.

It is at this point that Bell's Theorem comes into play, as these results provide us a way to test locality. Here's what we will do. We will allow the eastern and western beams to travel some significant distance from each other. We will then do our rough/smooth sort on the eastern beam and measure the western beam *at exactly same time as we make our sort of the eastern beam.* In this way, for the western beam to "know" that we have conducted a measurement on the eastern beam in time for it to play its little game and revert to half black and half white because we dared to make our measurement, would require such "knowledge" to pass between the two beams at a speed far greater than that of light! In other words, how could what I do to the eastern beam way over here, *instantly* affect the western beam way over there? It couldn't, not unless it violated locality.

What many predicted might happen was that the split beam (although comprised of one wave front of the same "thing") would "decouple" and become two things. This would maintain locality and show itself by the western beam remaining all white even though we sorted out the eastern beam. But that was not to be. As stunning as it might seem, when the eastern beam is sorted, the western beam will alter to half black and half white *instantly,* no matter how far away it is (or at least as far apart as nearly seven miles, which is the farthest it has ever been tested)! This apparent exchange of information violates the speed of light with a vengeance. Over the years, Bell's Theorem has been supported time and time again—quantum effects *are non-local*!

How could what's going on way over there, *instantly* affect what's going on all the way over here? It is as though there were no "over there," but rather that *everything* is "here" even though it might appear otherwise. I suppose, to a degree, that's what is meant by the effect being "local." If we apply the amazing results of the experiments supporting Bell's Theorem to the problems we face when assuming the existence of a "jumping" consciousness (so that we might overcome the supervenience paradox), we can eliminate the troublesome "jumping" aspect altogether. If everyplace is the same place, there is no need to "jump," indeed there is no *place* to jump. There is only "here." The mechanics of such a "stay in one place and share one consciousness with all people" might be a bit curious, but at least we would no longer be confronted with the problems associated with a mechanism for jumping a consciousness that we have already decided is generated by the organization of a brain that clearly would not be leaping from head to head.

At first glance, this might appear to be an awful stretch, and a clumsy way to solve the supervenience problem. I argue that it is neither. The reason for such incredulity probably lies in the assault that the proof of Bell's Theorem makes upon common sense. The proof shows us that it may very well be that there is no "there" there, but only a "here" here. In his book *The Whole Shebang*, Timothy Ferris introduces Bell's Theorem by stating:

> It may well turn out that over there—or, more properly, inside and underfoot, marbled through the very fabric of the space that is in turn marbled through every material object—the universe remains as it was in the beginning, when all places were one place, all times one time, and all things the same thing... [This] *might* mean that the universe is interconnected in some deep and as yet only dimly perceived way, on a level where time and space don't count.

How might the universe be interconnected in such a way that time and space wouldn't count (even though to our untrained eye they both seem quite important, if not fundamental)? There are many possibilities, and many of these possibilities are in the realm of metaphysics, as there is no obvious way to empirically prove their existence. For instance, physicist John Wheeler once thought that he might have an explanation for a very curious quality concerning electrons, namely that all electrons were interchangeable. By interchangeable, I mean that if you were conducting an experiment on a solitary electron, and for some reason it got lost (perhaps the electron broke loose from your containment apparatus and flew off to parts unknown) well, no matter, you can just use another electron. *They are all the same.*

The fact that electrons are all the same (although they might possess certain temporary characteristics such as "spin" that allow them to be

189

sorted into categories) causes them to be considered identical to one another, and for this reason also considered to be in some way "fundamental." The same is true of protons, neutrons, and many other particles. This is odd really; because when one looks carefully at any two seemingly identical things (snowflakes, fingerprints, diamonds, pebbles, etc) one will never find any two that are *exactly* alike. Fundamental particles appear to violate this common observation, which, if you stop and think about it, is pretty strange.

Wheeler once described to physicist Richard Feynman a concept he had that would explain why all electrons were identical. Wheeler argued that it was because *they were all the same electron!* In order to simplify Wheeler's argument it will be sufficient for our purposes to consider an illustrative example.

Wheeler imagined time, not as a straight flowing line the way that we typically perceive it, but rather as a great knot (there were good theoretical reasons for thinking of time in this manner as we shall see later). Since time and space are now viewed as integrated (thanks to Einstein), I want you to imagine the universe that we know as a slice through the space/time "knot." In this slice, all 4 dimensions, length, breadth, height, and time, are present. Surrounding this "slice" are higher dimensional hyperspaces into which quantum-sized particle/waves might travel, but into which molar sized objects (such as humans) may never venture (see Figure 23).

This one slice is frozen in time. Since it is "frozen" there is no change, and no change equals no time, so in this case time equals zero. Our one happy electron, however, is free to move about in the surrounding hyperspace. Each time it punctures the "slice," it will be perceived by the observer who lives within the slice to *exist* at the "puncture." Since the slice represents an instant frozen in time the electron (as it zips in and out—and appearing as a positron while traveling in the opposite direction) is able to "appear" in zillions of different locations *at the same time*....the same electron!

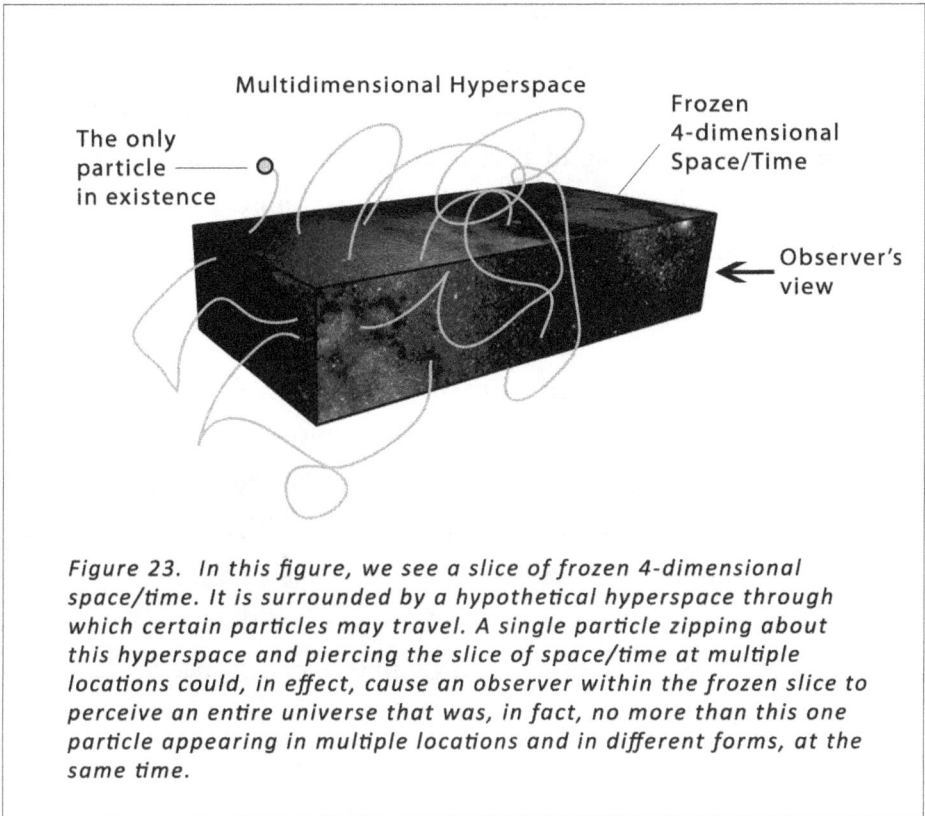

Figure 23. In this figure, we see a slice of frozen 4-dimensional space/time. It is surrounded by a hypothetical hyperspace through which certain particles may travel. A single particle zipping about this hyperspace and piercing the slice of space/time at multiple locations could, in effect, cause an observer within the frozen slice to perceive an entire universe that was, in fact, no more than this one particle appearing in multiple locations and in different forms, at the same time.

To be everywhere at once might seem quite strange, but you can begin to appreciate the curiosities that might associate themselves with various results when time is studied closely. Consider a photon that leaves a distant star and eventually strikes your eye after traveling for thousands of years. When the speed of light is reached, time stops. As you recall, Einstein imagined what it would be like to ride along with the photons reflected from the clock tower. As long as he was traveling with them at light speed, he would always see the time on the clock unchanged. The image was always the same. Because of this, photons don't know time. For them, time always stands still. From the point of view of a photon, it will have left the distant star and arrived at your eye at *the same time*. That puts the photon in at least two places at once. In Wheeler's multiple world line example of a space/time knot, the same fundamental particle can be everywhere at once owing to its multiple penetrations of a section of frozen space/time.

Of course, we are using an electron as the current example, but if such a thing were true, it would also refer to an even more fundamental "particle," the thing that is all things; the "thing" that the grand unified equation would be describing should such an equation ever come to pass. The whole huge and massive universe may be made of just one thing that threads itself through a single frozen "slice" of space/time. That one "thing" might actually be all that exists. The observer who looks upon the frozen slice and sees a great universe is, him- or herself, made of this one thing. That one thing, therefore, only exists "here." There is no "there." "There," is an illusion perceived by the consciousness created by the thing when it is "here." Consciousness can only exist when the thing is "here" to create it. And, "here" is where the thing will always exist because the thing is all that there is. It defines everything; it is the center and focus of all. Wherever it is, is "here." As an analogy, you might consider that if you took a very high speed photograph of the moving image you see on your television set, when you looked at the picture your camera captured you would never see anything more than a single illuminated dot. Amazingly, no more than one spot at a time is ever illuminated on your TV screen. The dot reappears an instant later over a tiny bit, and continues this process until it has covered a line across the screen. Then it makes another line, and another, until it completes one picture. A completed picture takes about $1/30^{th}$ of a second to construct. But at no time is there ever more than a single dot on the screen. The dot, at any given instant, is the entire TV show. The perception of a constant moving image that we see is an illusion. No complete image ever exists on the screen; not even a complete line. A single particle might be all there is, and as it weaves back and forth through a frozen section of space/time it might appear to us to actually be a complete universe made of many different things and yet, like the TV screen, there would never be anything but that one particle in existence; the rest is just that same "dot," appearing to be all places at once. If that dot represents all that there is, then "here" is wherever the dot is. "There" would only exist as an illusion.[*]

Interestingly, if a single "thing" or whatever you might wish to call it, is all that there is of the entire universe at any given moment, then one might rightly consider that the universe is not fantastically large, but rather, fantastically small. Actually, if a single "thing" is all that there is of the universe at any given moment then it actually makes no sense to even speak of size. Is the universe large or small? Well, large or small compared with what? If that one single thing is all that there is, then there is nothing with which to make a comparison and size becomes a moot issue.

For a better understanding of this possibility we will now turn to a man who lived over 2,000 years ago.

---

[*]Treating space as an illusion solves the split consciousness problem we earlier posed where one consciousness in placed on Mars, while its mate is left on Earth, causing perhaps a shared consciousness in mid-space. If all places are one place, then "Mars" and "Earth" and "consciousness" are also in that place as a single "thing." In other words, this particular paradox vanishes when our notion of space is collapsed.

ZENO OF ELEA

You can't get there from here.

—*Ogden Nash*

At this point you might justifiably throw up your hands and wonder what the heck I am talking about! Clearly there can't be only a single "thing" called "here" and nothing more. There must be both a "here" and a "there," otherwise motion would be impossible. After all, motion always moves you, or something at any rate, from here to there. If there were only a "here," there would be no place to go and motion couldn't occur. What would there be, then, that could go from one place to another if all things are "here?" Or is motion just an illusion? Well, that's an interesting question.

From where I am sitting I can look out of my home office window into my back yard. It's a pretty yard as yards go. There is a deck and a lot of flowers and a wooden picket fence a little over 100 feet away. For the sake of argument, let's say that the fence is 128 feet away. This observation now brings us to a man named Zeno.

Zeno was born in Elea, Italy, in or about 490 B.C. and lived until about 425 B.C. His philosophy, and especially his paradoxes, have puzzled and interested philosophers and mathematicians for centuries. Zeno's philosophy of monism was based on the idea that the many things that seem to exist are merely a single eternal reality. He called this single reality "Being." His primary concept was that "all is one" and that motion is impossible because there is no "there" there, since "here" is all that ever exists. Perhaps this is beginning to sound familiar.

It is an amazing idea for such a distant time, but Zeno went further and offered arguments to support his assertion. This brings us back to my picket fence, which is 128 feet away from where I am at this moment.

Zeno would ask that I start walking toward my fence, but with the following proviso: He would ask that I walk half the distance toward my fence, and then half again, and half again, and so on. If I follow his advice a very interesting thing happens. First I walk 64 feet, which is halfway to my fence. Then I proceed another 32 feet, half as far again (at this point you can see how lucky I am to have my fence exactly 128 feet away so as to avoid fractions). As I continue, following Zeno's instructions, I walk another 16 feet (never pausing, by the way). Then I walk 8 feet further, coming ever closer to my fence. Then I walk 4 feet, then 2 feet, then 1 foot. Now I walk half a foot, then a quarter of a foot, and then an eighth of a foot and so on. You get the picture.

As I proceed I make an interesting discovery. My discovery is this: Even though I will *never stop walking*, I will *never reach the fence*! In other words, although I move forever forward, I never reach my destination. I will continue on infinitely, and yet *never* arrive. How is such a thing possible?

Over the years, many people have studied this paradox and it has been dismissed by a considerable number as little more than a parlor riddle; nothing to take seriously since it can be easily refuted by simply walking toward any such "impossible to reach" object and laying a hand upon it. However, when mathematicians examine the paradox carefully, the issue remains far from clear. Consider that there are an infinite number of points in space between my fence and me. It is reasonable to assume that it would take some time, however small an amount, for me to move from one point to another. But because there are an infinite number of points, this would mean that it should take me an infinite amount of time to reach my fence. An infinite amount of time is an amount of time that would never end, ergo; I should never reach my fence. But I do reach my fence. So what is going on? *

If Zeno is correct, and there is no "there" there, but all is Being, that is, all is "here," then perhaps when I "walk to my fence" I don't actually "move," but rather what is "here" changes in some way to produce my fence "here" when it used to not be "here."

If all that seems a bit hard to swallow, let me come to the same conclusion from a different direction, a direction that should be clearer to you because it will be a bit more familiar. One way to overcome Zeno's paradox is to make a particular assumption about space, and indirectly, about time as well. Physicists have shown that energy is released in packets that are called quanta. A photon is one such packet, or quanta. It travels through space as a compact unit of specific energy. Imagine then, if you will, that space also comes in units, or quanta, and that there is a space so small, that no smaller space could exist. This puts a bottom limit on how small space can be. The same can be said for time. There might be a time that is the shortest possible. Many physicists believe that this is the case and that the smallest units of length and time can be defined by combining Newton's gravitational constant and Planck's quantum constant. Numerically, these are given as 10^{-33} centimeters for the smallest possible length and 10^{-43} seconds for the shortest possible time. If this were the case, then I could not continuously move half again as close to my fence with each step, because sooner or later I would be required to make a step smaller than 10^{-33} centimeters (now that's what I call a baby step), and there would be no space available that was small enough to accommodate it. If space

---

* Some mathematicians have argued that calculus can solve Zeno's Paradox inasmuch as when one begins to place an infinite number of points along a line, the amount of time it takes to go from point to point approaches zero, allowing one to possibly pass an infinite number of points in a finite time. Yet many take issue with this solution.

exists as quanta, then I could reach my fence. Since I do reach my fence, and since much that we find at the very smallest levels do come in quanta, it is tempting to believe that space itself, comes in specific units of extraordinary small quanta. In other words, space is digital and not analog!

Considering how a motion picture works can provide us with the best analogy of such a world. Everyone knows that when he or she watches a movie, that there is no real motion on the screen. The apparent motion is an illusion caused by rapidly projecting many still, or frozen images, on the screen in quick succession. That's all a movie is—lots of still images, frozen in time, shown rapidly one after the other. Although we clearly sense smooth motion, no such motion actually exists.

Perhaps the real world is much the same. As I walk toward my fence, I do not move smoothly toward it, but rather I exist in trillions of frozen bits of individual space/time that I experience in rapid succession, and which yield to my mind the strong illusion that I am moving freely and continuously through a non-segmented space. If my movement were not continuous, but rather a series of frozen moments in space/time run together, that would explain how I was able to pass through a supposed infinite number of points on my way to the fence without requiring an infinite amount of time in which to do so. In fact, if a motion picture were a continuous stream of analog data and not "digitized" by being organized into a set number of frozen frames, it too would be infinite in length.

Our daily existence could very easily be comprised of a series of frozen moments through space and time—frozen moments of "here." If the moments were short enough, or in the parlance of the movie business, the flicker frequency or frames per second, were high enough, we'd be none the wiser. We'd simply go about our lives believing that we were smoothly moving from place to place. Many modern physicists believe that a series of frozen frames is the only possible way that "motion," or at least motion as we perceive it, could occur.

THE FROZEN UNIVERSE

Darest thou now, O Soul,
Walk out with me toward the Unknown Region,
Where neither ground is for the feet, nor any path to follow?
No map, there, nor guide,
Nor voice sounding, nor touch of human hand,
Nor face with blooming flesh, nor lips, nor eyes, are in that land.
I know it not, O Soul;
Nor dost thou—all is a blank before us;
All waits, undream'd of, in that region—that inaccessible land
—Walt Whitman, "Leaves of Grass"

Things really begin to become interesting if we stop to consider what one of those frozen moments in space might actually be. Imagine that you are walking toward a fence of your own choosing (I'm kind of private about my yard). Imagine also that your "smooth motion" toward it is, in reality, a series of frozen "frames." Let's stop for a moment and consider just one of those frames.

Let us imagine that the *frame*, as I shall continue to call it, is a frozen moment in space/time. It is an instant in your life when you were approaching a fence. Everything is frozen absolutely still. Your conscious mind, your perceptions, your memories, are all frozen in that instant. In short order you will have jumped on to the next frame. It is a quantum jump of sorts, inasmuch as you got closer to the fence without having to pass through any space between frames. You are moving just like the actors in a movie, jump, jump, jump, from one frame to the next. Your "motion" is a series of "quantum leaps" from one frozen frame to another, thus overcoming Zeno's paradox and enabling you to eventually reach the fence. These frames are so numerous, a fantastic number per second, that you are quite unaware of their presence.

Now, imagine that you are sitting in your living room, in your favorite chair. You will now rise and walk across the room. We can represent your journey as a series of trillions of frozen frames. For the sake of our example, let's just select 10 of those frames— the first, the last, and eight in between. That will be enough for our purposes. Let us call you person "A" and we shall label these frames that have you in them A1, A2, A3, A4, A5, A6, A7, A8, A9, and A10. In frame A1 you are seated. By frame A10 you are on the other side of the room.

Let's start with the slice of frozen space/time we are calling frame A1, as that seems like a reasonable place to begin. Now imagine that A1 is *all there is?* Suppose A1 is the entire universe? I mean—the works! I suppose that the universe wouldn't be very big, then. There isn't a whole lot in A1. Think about it. A1 contains your conscious mind, whatever you happen to be perceiving that very instant and perhaps whichever memory you might be in the process of accessing, but nothing else. In this view, a "frame" is the same as a universe and, like the frames of a movie, as one universe or frame ends it is replaced by a different one in a perhaps never ending sequence.

Perhaps the entire universe of frame A1 consists of only what I have listed. There are no Moon, no stars (unless in frame A1 you happened to be looking at one), nothing but your mind and what you might be experiencing at that very instant. As Descartes said, all he could know for certain was that he existed.

If A1 were all that there was, if it were the *entire universe*, then the supervenience paradox, Zeno's paradox, the proof of Bell's Theorem...none of these would any longer be a problem. They would all make perfectly good sense. The supervenience paradox is no longer a problem because only one consciousness exists in *the entire universe*, namely yours. Zeno's paradox presents no concern because there is no motion in A1; all is frozen. The proof of Bell's Theorem now makes sense—there is no *there*, only a *here*, at the exact point of consciousness that exists in frame A1 (i.e., all apparent non-local effects are, in fact, really local). The idea that you are the only mind in existence is known as solipsism, from the Latin *solus*, which means alone, and *ipse*, which refers to one's self.

Curiously, solipsism is a position that can be derived from quantum theory, a view with which Einstein took issue. When dealing with this concept, Einstein used to pose the question, "Is the Moon there if you are not looking at it?" The argument encompassed by that question was that there is a "reality" apart from any observer. In other words, Einstein argued that yes, the Moon is there, even if no one is looking at it. But quantum experiments, especially those dealing with non-local effects, appear to suggest that nothing actually "exists" until it is observed (somehow, it was assumed, causing the wave front to collapse and produce a "real" object). Slide this interesting debate onto a back burner for a moment and we will return to it shortly. In the meantime, let's go back to our frames.

Let us assume that the *entire universe* in frame A1 is limited to the few items I suggested. Next, universe A1 is replaced by universe A2 (don't worry about why A2 should appear instead of something totally different—we will get to that). A2 is only slightly dissimilar. A2 also contains your consciousness, but your perceptions are a bit different. You are standing up now, and perhaps you're thinking of something else now that you have gotten off of the chair. The universe in frame A10 is different still, but not a whole lot different. It still contains your conscious mind, but now you have crossed the room. Ten tiny universes all in a row; none of them has a Moon in them since you weren't looking at a Moon in any of the frames. So those 10 universes are moonless.

Let's play another game with these 10 frames. Let's put them in random order. I will sit here at my desk and roll dice to determine the order. (I actually have a 10-sided die my son uses for a role-playing game). By rolling this die I will create a random order for my 10 universes. I might add that since Einstein said, "God does not roll dice," I can now be acquitted of any charges of "playing God." I have come up with A1, A5, A4, A9, A6, A2, A8, A10, A7, A3. (This was not totally random, since in order to help our example along I left out repeats). It seems that we are starting with A1 again, well...when you ask for random, you get random. Imagine mixing up the frames of a motion picture in this way. Wow, you'd end up with an unwatchable jumble of images. Let's see what we get when we scramble up our frozen universes.

In universe (or frame) A1, you are sitting in your chair (again). But now, you suddenly jump to frame A5. Would the "you" in frame A5 notice that you had skipped frames A2-A4 and suddenly leapt halfway across the room? If you stop and consider what frame A5 contains, I think it will become clear that you would be blissfully unaware that you had leapt anywhere. In frame A5 we have a few things. There is your conscious awareness, there are the perceptions that you have about the room based upon the sensory inputs you have received, and there are the memories that you might be actively retrieving at that instant. The memory that you will possess in frame A5, will be one of *having passed through frames A1 through A4*, since frame A5 has, by definition, that memory included in it, indeed, the possible memories you will discover that you might have in frame A5 could be any from your life up to that very instant. This is what frame (universe) A5 has in it...your conscious mind and a sample from your life's memories, and what few things you perceive at that very instant. You might well ask, how does frame A5 know what happened to you before? The answer is that it doesn't! Any universe that had your conscious mind in it and also included memories of frames A1-A4 (even if such frames had never existed) would be defined as FRAME A5.

If this is unclear, consider that frame A5 is a moment frozen in time. It has no past or future. It is only of the present. Time stands still in frame A5. Since time is change, we can say that in frame A5, time does not exist.

Because all time has ceased to exist (or, if you wish, you can think of the future and past as having collapsed onto the present that is contained within an individual frame), any memory you have is not from the past, but is simply a memory that exists now, a thing of the present. If you think about it, all memories are *present* events and thoughts of a "past" can only occur in the present. The "you" in frame A5 might well have memories of having gotten off of the chair and walked halfway across the room, but that doesn't mean that such a thing ever actually happened. But we are calling this frame A5 because it does contain those "A5 memories" along with your consciousness.

This is not unlike the situation earlier when we tried to imagine an improbable "leaping" consciousness that might go from one body to another. If your consciousness jumped into the body of another person, you would recall all of that person's past and assume that it was your own past (since that person's memory was the only one that could combine with that particular brain to generate your consciousness). You would think that you had done all the things that that person had done, even though "you" in fact, weren't there during those times.

Perhaps another example will make it clearer that present memories determine the sort of past we think we have led. Imagine that you are a person living in the 28th century and you wish to know what it was like to live a typical life in the early 21st century. For a couple of hours your brain is wired up to a great computer. Next, the computer artificially inputs sensations that mimic reality as it was in the 21st century. It also blocks your real memories, and instead provides you with a full life history should your brain attempt to access memories of the past in an attempt to confirm who you are. The computer assigns a number of appropriate 21st century variables about your life at random (such as what you might be doing, or wearing, or how you will look, where you will live, etc.). The computer then does a check before it allows the experience to continue. In fact, I am that computer and this is that check. Do you think you are alive in the 21st century on the planet Earth? Yes? Do you believe that you have been you for many years and have a past that is in accordance with the person that you think you are? Yes? Do you know anything about the 28th century? No? Does the term *natarnl akkat passal* mean anything to you? No? Very good, then the program begun only 4 minutes ago is in place and you may now experience a few days of typical life in the 21st century on Earth. Enjoy!

How do you know that it isn't true? How can you know? You can't know. There is nothing more to be said about it. Just because you have memories of having done something or of having been someplace doesn't mean that you did that thing, or that you were ever in that place, or even that such a place exists. Also, just because you can perceive photographs of those times and places doesn't mean that there is some sort of "real" record of the past. Those, too, are merely present perceptions. For this reason, you could have memories of moving to frame A5 from A1 even if you began the sequence with frame A5.

Look at Figure 24 and you will see the 10 frames in two different arrangements. In one arrangement, they are continuous and sequential. Time flows right along, straight as an arrow. In the other arrangement, the timeline is haphazard and random, even passing through some of the same frames more than once. Those frames are in random order. Unlike the scrambling of the frames of a movie, which would leave the audience totally confused, if the consciousness resides *within* the frames and draws on the memories contained within each frame, there would be no perceptible difference between the frames in a linear or a random order. You can scramble frames A1 through A10 all you like, and the consciousness experiencing them would be none the wiser. His or her experience would be the same whether the frames were in sequence or not.

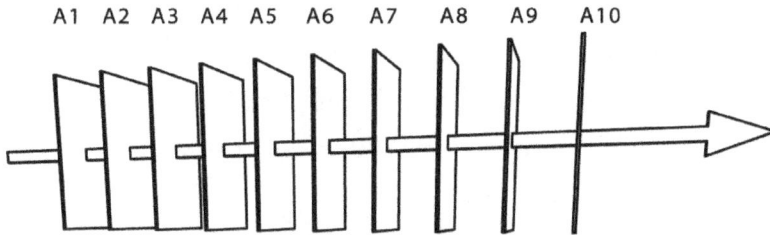

The flow of time progresses sequentially.

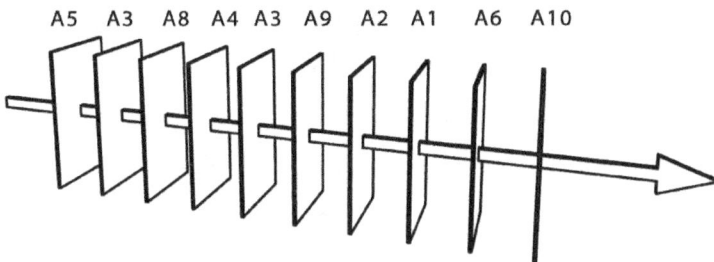

The flow of time progresses randomly.

Figure 24. Two arrangements of 10 frames of frozen space/time (arrow): Each frame represents the arrangement of the entire universe at any given moment.

The top series is sequential and follows the arrow of time. Our perceptions lead us to assume that time flows in such a sequential manner.

The bottom arrangement shows the 10 frames of frozen space/time in a random order. If time flows in a random fashion, the "future" e.g., frame A5, may appear before the "past" i.e., before frame A3. Since each of the frames contains a consciousness with access to a memory that is self-contained, an individual could "progress" from frame A5 to A3 without ever having the sense of going back in time, because frame A3 does not hold memories of frame A5. A person in frame A5 would have memories of frame A5, as well as memories of Frames A1, A2, A3, and A4. The presence of such memories would define frame A5, whether or not frames A1-A4 ever actually existed. This leads to an illusion that time flows forward in a sequential manner when, in fact, the flow of time might be completely random.

He or she would rise from the couch and walk across the room. Each frame is internally consistent. The memories in each frame are of a life up to that "point." Backing up from frame A5 to A3, for instance, causes not the slightest surprise, as frame A3 contains a consciousness with access to no memory of frame A5 (if you don't know that you have been to the future, you can't be surprised to find yourself back in the present). All the frames of your life, the trillions and trillions of frames from your birth to your death could all be "passing" in random order, a wild scramble filled with repeat frames and, frankly, you would never know it. Such a scrambled life would be indistinguishable from one in which all the frames passed by in rigid order. How time flows doesn't really matter.

Physicists have often wondered why the arrow of time points forward. Well, who says that it does? The flow of time from past to future that we all seem to experience and take for granted might be nothing more than an illusion. In fact, time itself might be a grand illusion; it might not even exist.

WHO'S GOT THE TIME?

Time is what keeps everything from happening all at once.
Space is what keeps everything from happening to you.
—*Neill D. Hicks*

I grant you that time, when examined closely, is a very strange thing. Still, time seems so much a part of our lives that we typically take its existence for granted and rarely consider what it is, or even *if* it is.

Giving time short shrift, however, can leave us wearing logical blinders. For example, there is a delightful conundrum that goes as follows: If God is all-powerful; can he make a rock that is so heavy that even he can't lift it? At first glance it would appear that God has gotten himself into quite a jam on this one. If he *can* make a rock that he can't lift, then he can't be all-powerful because he can't lift it. And, if he can't make a rock that heavy, well then, he can't be all-powerful because there are clearly limits on what he can create. It appears, then, that God can't be all-powerful. This little conundrum when brought forth at the wrong time and in the wrong company has led to the loss of the occasional head, but in general, it is a good party stumper.

Of course, *time* is the key element here. To solve this problem, God need only suspend time. Then he can make a rock that is so heavy that he can't lift it *and* (at the same time) lift it. Furthermore, God gets to do all this because he *is* all-powerful.* Time, as I said, is a funny thing to play with. Some physicists, to the shock of others, have even argued that time doesn't exist.

---

* Here is another fun one. Is God omniscient, that is, is he all-knowing? If so, then He must know who is going to Heaven and who is going to Hell before that person is ever created. So how can people be held responsible for their actions? Free will you say? Well then, God doesn't know what will happen and he is not omniscient. Such arguments have split churches one from another.

One of the current champions of this view is Julian Barbour, a theoretical physicist. In Barbour's view, time does not exist. In his book, *The End of Time*, Barbour paints a picture of a universe in which time plays no role. There is no past or future. The entire universe, all the trillions upon trillions or stars and planets, including our own world and all the people on it, are frozen in a great still-frame. This one great frame *is* our entire universe, and nothing in our universe ever changes.

This idea is not a new one. In 1949, the philosopher and mathematician Kurt Gödel made an amazing discovery. He demonstrated that there could be other universes. Based on Einstein's theory of relativity, Gödel referred to them as "rotating universes," in which time would not exist. In such universes, it would be possible to travel both forward and backward in time. By being able to travel backward in time, Gödel argued, the "past" would never be past, but always present. The same could be said of the "future." If all were "present," then time is frozen, or to put it another way, since change is time, time would not exist in a universe in which all was "the present."

Gödel then conjectured that our own universe might be such a universe. Gödel worked at the Princeton Institute for Advanced Study and he and Einstein walked home together nearly every day for many years. This gave Gödel an excellent chance to discuss his views with the founder of relativity theory. Einstein's reaction to Gödel's proposal was to admit that it was a remarkable discovery, and that it opened up many disturbing possibilities concerning relativity. However, Einstein died before he could fully address the issue. If, as Gödel asserted, in our universe it is possible to travel forward and backward in time, then relativity theory doesn't just explain time; it explains it away!

Others have also explored this possibility prior to Barbour making his argument. In their seminal work, American physicists Bryce DeWitt and John Wheeler pointed out that the universe, by definition, has nothing with which to interact other than itself. During all of the interactions that might occur within the universe, the net amount of energy never changes. Energy is neither created nor destroyed. As you will recall, even when matter seems to appear out of "empty" space, gravity, a negative energy, appears in a sufficient amount to maintain the total amount of energy equal to what it was before. Something is not created out of nothing, because after the "creation" it is still, essentially, nothing. DeWitt and Wheeler argued that if the overall change in the universe is always zero, and since time is change, then it might make good sense to drop time as a variable when one endeavors to develop an equation that attempts to describe the entire universe. If one wishes to make progress, like God with his heavy rock problem, it might be best to drop time out of the picture. Stephen Hawking and other eminent physicists take DeWitt and Wheeler's work seriously, although not all believe that a grand theory could describe the real universe without the inclusion of time.

But Barbour and others persist, arguing that every potential arrangement of our universe, including what we call the past, present, and future, actually exists, individually and eternally, but frozen. Each separate universe is a "now." Barbour refers to all these universes collectively as *Platonia,* after Plato, who argued that reality was composed of everlasting and unchanging forms.

In this analysis, each universe is a frame, but a frame that will forever stand alone from all other frames. There is no interconnection, no timeline. There are just endless *nows* all different from one another perhaps owing to random chance. Perhaps some frames are identical, but that would be of no consequence, since no two frames ever interact. In one frame you might be an old person, in another a child, in another middle-aged. If this sounds familiar, it should, in this instance we are reiterating the idea of frames A1 through A10. In that example, there were ten *nows,* and they need not be related to one another. They can each be considered to be a separate universe.

A FRAME BY ANY OTHER NAME

There is no fear of the cycle of birth, life and death.
For when you stand in the present moment, you are timeless.
—Rodney Yee

What's actually in a frame? Let us look very closely at what we might suppose it to be. First, is it large, gigantic? Is it filled with galaxies, stars, and planets frozen in time? Is the Moon in the frame even if you are not looking at it? If the Moon is there even though you're not looking at it, if it and everything else is "real" in the typical way that people mean "real," then all the people on the Earth will be real and will also be in the frame. If so, then they are present, along with you, frozen in that particular "now" that is the frame and I dare say that this will, once again, raise the supervenience paradox. But perhaps all that stuff isn't in a frame. Perhaps it is not as complex as that. In fact, this might be a good time to apply Occam's Razor. Occam's Razor was named for William of Occam, a 14th century English philosopher who argued that, "entities must not be multiplied beyond what is necessary." In other words, if there are many ways to describe a phenomenon that appear to match one's observation, choose the simpler description.

Let's start with the most basic and simple of assumptions. Let us say that these frames, or independent universes, come into being because absolute nothingness is unstable for whatever reason. These frames pop into existence and contain, well, I don't know what they contain. But let's assume that each frame has some physical laws that it follows, even if these laws came about in a totally random and arbitrary way. We have discussed such a concept before when we wondered if the laws of physics might only be "laws" as such in this universe, while in other universes different "laws" appeared, all owing to random chance.

Apparently, some of these "nows" or frames, contain a little patch of consciousness or you wouldn't be reading this book. I am guessing, if we are talking about a random process, that most frames will contain no patch of consciousness. Why should they? If we are talking about the random popping up of frames, each following random physical "laws," and generating various internal "sparks" then I would guess that a patch of consciousness would be an exceedingly rare thing for such random "sparks" to produce.* Then again, if you are talking about an *infinite* number of frames, then patches of consciousness would be in abundance so long as it was ever possible for one to occur in the first place. In fact, there would be an *infinite* number of them in the same way that there would be an infinite number of integers divisible by 100 in the series "1, 2, 3, 4, 5, ..., infinity."

Before going further in consideration of what a frame, or "now," might contain, let's take a moment to get a clearer idea about what a "now" actually is. Time is so obvious to us all that it is very difficult to fully grasp a "now."

Take a moment and try something. Touch the tip of your nose with your index finger, and then put your hand back where it was. Next, think about the feeling you had when you felt your finger touch your nose. Now, *at this very instant*, let's freeze time. Imagine that it has frozen right at the moment I asked you to think about how it felt to touch your nose. The question now is, who, or what, are you in the context of this frozen frame?

---

* In this one instance, I am breaking with the philosophy of solipsism, in that I argue that such frames can exist in an "empty" state, that is, without containing a consciousness. Of course, there is no way to ever know if this is so.

It is reasonable to assume that you are a patch of consciousness frozen within this universe, or frame. There is the sense you have of being aware that you have a self. You also have a memory of touching your finger to your nose. There is no Moon in this universe, no stars. You have no parents, no previous life because you weren't thinking of them. There is a *memory* of touching your nose, but I would argue that it is likely that you never actually did touch your nose. In fact, I am not sure that you even have a nose. There is a bit of flesh-colored blur out of the corner of your vision. Is that a nose? Who can say what it is? The rest of this universe, or "now," is comprised of the few other things you can sense, *and that's all*. And it will stay like that, frozen forever and ever.

Perhaps you can't believe it. After all, you are moving through time, you can sense it. I, too, can look out of my window and see the leaves on the trees rustling in the wind. But if I think only of the *present*, the *now*, I will know that the only image I could possibly be seeing at any given instant is a tree frozen *in the present*. The motion of the leaves is a *present memory* and not something that I am actually seeing. Perhaps the leaves have never moved, but I simply think that they have because my memory tells me so.

I have seen many fine shows on television and I remember many of them. But at no time was I ever looking at more than a dot on my TV screen. The present that you sense right NOW might be all that there is, forever frozen and eternal. There is only a little patch of consciousness that senses the now and the few memories that it can access. The feeling that you have lived through thousands of "nows" is only a memory that exists in the now. If this is true, there is no afterlife, there is only one life and we are immortal. You might be experiencing a frame that is right at the moment of your death, but you could never actually die.

The idea of frames is a very useful one as far as physics and philosophy go. As was noted, if we are the only patch of consciousness in a frame,

213

then the troublesome problems brought on by supervenience, Zeno, and Bell's Theorem disappear. But by totally dropping time out of the picture we create another dilemma. If each of these frames is a universe separate and unto itself, then how is thought possible? Descartes said, "I think, therefore I am." Thought requires a certain amount of processing, and so does the retrieval of a memory. Anyone working with a computer knows that to operate the central processing unit and make any sort of calculation requires time, however little. How could you be conscious if there were no time in which to process the information that would lead you to become self-aware?

At this point, a good philosopher might take note of a nasty problem in my reasoning. I am now relying on observations of things and experiences that might not be real in order to preclude the possibility that something could or could not happen in a universe that I could never hope to understand and which is totally foreign to me since I am but a frozen patch of conscious awareness residing within it created by forces that will forever remain a mystery to me (ugly, isn't it?) At this point one is tempted to put down a bowl of cream for the Cheshire cat and simply walk away. It just seems like the right thing to do.

But never one to say die, I will press on. Perhaps the flow of time is an illusion brought about by the existence of all these independent frames, but I am still tempted to ask how, if there is no interaction between these patches of consciousness as Barbour suggests, is the illusion of time passing, or even the illusion of being aware, at all possible? To be possible, thought and memory in this frozen world must operate without biochemical activity (which requires time). Of course, at this point I am facing something of a tautology. I am assuming that time is required for consciousness, while it might well be the other way around. Perhaps in some nearly magical way, the presence of Platonia generates some vast sense of gestalt or unity among all the patches of consciousness even though these universes remain forever unconnected. It's not up to me to say no. But if that is what is happening, it is far beyond the boundaries of logic and way out of my reach. In fact, with a nod to Niels Bohr, it's so whacked that it's probably exactly what's going on. If it is what is happening then I will say the dreaded phrase that I hoped never to hear...then anything's possible...and hit myself in the face with a pie.

Sometimes it is difficult to imagine that there is a debate among some physicists as to whether there is such a thing as "time," because time itself seems so basic, so fundamental. Perhaps time does not exist in some universes, but does exist in others, simply as a matter of random chance as to which laws of physics abide in any given universe. Perhaps our own universe does contain time, and is not like the ones that Gödel showed were possible. One way to demonstrate this would be to show that time travel in our universe was not possible, and that the past and the future are not part of the present. So far, it has been proven that it is possible to travel into the future. In fact, any time you add to your velocity, time slows down for you in your own frame of reference. By driving a car (or even walking) you will age a tiny bit slower, or conversely, the things around you that are moving at a relatively slower speed will go forward in time faster—in essence causing you to have moved into the future. At low speeds the changes are so subtle that we do not notice them. At higher speeds it becomes more apparent. Astronauts returning from the Moon were found to be two full seconds younger than if they had stayed on Earth. From their perspective, they had gone 2 seconds into the future. Some jokes were made at NASA about docking their pay since their mission was shorter than their contract had called for.

Whenever satellites are placed in orbit, engineers know that they must build in a time correction for the effects of relativity. Orbiting satellites travel at roughly 17,500 MPH and, as a result, time flows slower for them than it does for us on Earth.

At extremely high speeds the time differences would be very noticeable. The famous example often given is of twins. One twin stays on earth while the other rockets off into space and travels for a few years at nearly light speed. Upon return to Earth, the traveling twin will discover that he is now a few years younger than his sibling! From his point of view, he will have traveled into the future a few years in what may have seemed to him like only a few weeks time.

But such proofs are insufficient to allow us to say that our universe has no time. To do that one would have to show that it was also possible to travel *back in time*. Going backward, however, can lead to all sorts of trouble. There are many fictional accounts of time travel, but only a few have attempted to address the real dilemmas that would emerge if one traveled backwards in time. The most famous is known as the grandfather paradox. Stated simply, imagine that the time traveler goes back in time and kills his own grandfather before the poor fellow has had a chance to have any children (the reasons for this brutal act are rarely given). Once his grandfather is dead there is no way that he, the time traveler, could ever have been born and, by extension, no way for him to be there to kill his grandfather in the first place.

In consideration of such problems, Stephen Hawking has argued that it is most likely not possible to travel back in time, at least not in our universe. He proposed this with his now famous chronology-protection conjecture. Hawking began by asking a simple enough question, if time travel were possible, he asked, then why have we not been invaded by hordes of tourists from the future? Or, better yet, why isn't our history filled with stories such as, "John Wilkes Booth attempted to assassinate President Lincoln on April 12, 1865 at Ford's Theater, but two men in strange clothing suddenly appeared and wrested the gun away from him. Before these strangers could be identified and thanked, they appear to have left the scene."?

In response, some have argued that perhaps a new universe would be created as soon as someone altered the timeline (one universe with a healthy grandfather and one with a dead one). Or perhaps it is not possible to travel back in time to a period before time machines are invented. Or, perhaps, as Hawking argues, in our universe it is not possible.

Physicist and astronomer Kip Thorne has examined a number of possible solutions to the time travel problem by imagining a wormhole in space (essentially a distortion of space-time that would link one location or time with another, through a path that would be shorter than would otherwise be expected). Through the use of such a wormhole, Thorne showed that it might be possible for a spaceship to safely travel through a very short wormhole (certain types would have energy distributed in such a way as to allow for a safe passage) and emerge in time to see itself entering the other end! Imagining a billiard table with a small wormhole that allowed the balls entering it to travel back in time and appear in the billiard shot at an earlier moment, yielded one solution that showed a billiard ball exiting the wormhole and colliding with its earlier self! Interestingly, though, the collision of itself with itself knocked itself into the opening of the wormhole where it had just gone before! An example given from that solution involving a time traveler is of a soldier who fights in a battle and is wounded in the shoulder. After receiving the wound, the soldier's aim is permanently affected and he can no longer shoot well. Later, the soldier realizes that he was on the wrong side during the war and uses a time machine to go back and make amends by fighting for the right side this time. As he fights, he comes across an enemy soldier and fires at him. However, because his aim is bad owing to his old injury, he only hits the enemy in the shoulder. Of course by now you have guessed that the enemy soldier he has shot is himself, thus causing the original injury. This solution allows for the future to affect the past. Such a solution, if it applies to our universe, might well mean that events are predestined. Even so, our universe would be one in which time existed.

If we accept the point of view that time does exist, then each of the frames that we have been discussing can be considered as a special quantum within a single universe through which a timeline runs, so that the frames pass by like the frames of a motion picture. (Or, if they are not "passing by," new frames are being created each instant much in the same way as virtual particles are constantly being created, annihilated, and recreated). The only difference is (as we discussed before) that the timeline need not follow any particular order. Regardless of which order the frames come, life will appear to follow a straightforward chronological line to the consciousness residing in them because of the way in which memory functions.

My assumption at this point, then, is that for a consciousness to be generated by whatever forces are present within this universe, or frame, a certain amount of memory and perceptual information is required. If the right bits of this thing we call memory (I am not talking about brain cells now, but rather some force in the frame that we might never understand) mix with the right bits of what we refer to as sensation and perception, then a consciousness can be generated. In most frames, the "sparks" (for lack of a better term) within the frame yield no consciousness. But, if just the right pattern of "sparks" appears, they yield a consciousness that most likely has some thoughts, memories or perceptions as required components, because without the "thoughts," "memories" or "perceptions" I conjecture that the consciousness wouldn't be generated as there wouldn't be much to be conscious about.

Perhaps another good reason to consider that our universe does have time in it, and that the physical laws of our universe require that time be considered, is that we do not have memories of the future. If we did live in a universe without time, and a memory could be accessed by a conscious mind without the need for time, and that memory were a random thing, then why wouldn't it be just as likely that we would have memories of the "future" as well as some from the "past?" Yet, none of us has memories of our self when we were older than we are now.

If a certain strip of frames randomly tossed together has the power to generate a consciousness as well as some memories associated with that consciousness, but the memories are *always* of past events, we might assume that such a frame can only exist in our universe because certain antecedent events *could* have led to its present organization (whether such events actually occurred would not matter). It is a difficult concept to explain, but just think of any frame, appearing randomly (and having the laws of physics with which we are familiar, thereby making it, by definition, a frame from *our* universe) in such a manner that its organization describes a *possible* past that would have been lawful, in terms of the physics of the frame. If a timeline ran through a series of these random frames, and the organization of these frames generated a consciousness, no memory accessed (whether the memory was of a "real" event or not would not matter) would ever appear to be a memory of the "future."

Of course, we are once again courting a tautology. Perhaps the current frame you are in just so happens to only have a memory of a past event, and that the next frame to come along will have a memory of you when you *were* older than you are now. We can't logically argue that no one has memories of when he or she was older, if we ourselves are stuck in a frozen frame that, by random chance, tells us that such a particular supposition is true. Perhaps the next frame, just as randomly, will say something different.

Even with time placed back in the equation, the situation is far stranger than you might think. Let's start with the same basic assumptions as before. Let us say that these frames, contained within one universe, come into being because absolute nothingness is unstable for whatever reason.

219

Again, the frames pop into existence and contain who knows what. But let's assume that in our universe, at least, all the frames follow the same physical laws, even if these laws came about originally in a totally random and arbitrary way. We will consider frames that follow different laws to be rightly considered belonging to a different universe (one that operates according to those laws of physics).

Once again, many of these frames contain no patch of consciousness, but if there are an infinite number of frames, then a lesser, although infinite number of frames, do contain a consciousness. (Hard as it is to grasp, there really are larger and smaller infinities.) There is no guarantee that there are an infinite number of frames, of course, but we can surmise that for a conscious thought or memory to exist, there must be trillions of frames with the same consciousness and slight variations of the same thought or memory that can connect together as a unified entity. As an analogy, imagine that I told you that I flashed a bunch of random dots of light onto a TV screen (each dot being likened to a frame for purposes of our analogy). If I flash the random dots all over the screen, and quickly enough, the entire screen will seem to turn on, but you would see no picture, you'd just see what is commonly referred to as "snow." However, if I flash random dots long enough (and this is not unlike the old idea that if enough monkeys hit enough keys on typewriters placed before them, that sooner or later one of them will compose Hamlet), it is possible that a scene from "The Lord of the Rings" will play out. In our analogy, this would be akin to a burst of conscious awareness. Of course, you wouldn't know how many dots I would have to *randomly* flash onto the screen to get such a result, but to eventually hit that sort of a sequence with random dots it must have been a lot of them to say the least. For a series to exist among the zillions of possible frames sufficient to yield a conscious thought or memory there must be a lot of frames out there. Maybe not an infinite number, but something close to it.

Recalling our old example of standing up from the chair and walking across the room, we can now say that scattered among frames A1 through A10 there are many frames without a patch of consciousness within them. Think of them as "empty frames," if you like, since there is no consciousness in the frame to perceive whatever else might be in there. Just like with the random dots on the TV screen, in a very few rare instances we get a picture, but in most cases we just get "snow."

TV sets follow the laws of physics, and I am going to assume that the frames within our universe follow the laws that came to be in *our* universe. Here, I am defining *our* universe as any frames that follow the same laws of physics. Frames with other laws that are strange to us, I am calling rightfully parts of another universe. But this is nothing more than a convenience. I am just using it to help us along in our thinking. In the frames of our universe, certain things can happen in them and certain things cannot. This might explain why you have no recollection of having a green tentacle in place of your left arm 5 minutes ago. It might also explain why we have the "illusion" that the sun rises in the east and sets in the west every time, and why we have the "illusion" that the theory of evolution appears to hold up so well. I use the word "illusion" in quotes to denote that the memories or perceptions allowed to accompany the generation of consciousness in our universe will appear to support certain lawful scientific relationships.

This might seem like quite an assumption, but if it is not the case, and if the memories associated with each patch of consciousness were totally random things, our experiences might not appear to be very logical or consistent. So I assume that there are some laws that are in command concerning what sort of memories are allowed, and that they need to be consistent with the laws of physics. Of course, who's to say that consistency isn't in the eye of the beholder and that if you did have a green tentacle for a left arm that it might not bother you so long as you also believed that it was normal to have one. Still, an entirely random process of matching memory, perception, and consciousness would, I think, lead to most consciousnesses containing frames holding minds that were undergoing extreme psychotic experiences that made no sense whatsoever because memory, senses, and understanding of what it all meant, were totally mismatched. Perhaps it is possible that at this very NOW you happen to be in a "rational" frame, rather than one in which you are aware that you are a washing machine with the bill of a duck who is attempting to keep from choking on blue glue. Even that would be sensible compared with most random frames, since each aspect, a washing machine, a duck, the color blue, and glue are internally consistent. Most random frames would hold a consciousness that was clueless, living in a "world" that meant nothing and was senseless. Perhaps most frames are like that. I hope not!

I think it makes more sense to believe that within each frame in our universe the laws of physics reign and that certain logically rational memories, sensations, and perceptions are allowed to accompany a patch of consciousness, while others are not. This issue is a complex one, however, and we will examine it more closely in the next few pages. As you will discover, it leads in an amazing direction.

THE WORLD AS WE KNOW IT

Another one o' them new worlds.
No beer, no women, no pool parlors, nothing.
Nothing to do but throw rocks at tin cans.
And we gotta bring our own tin cans.
—*Cyril Hume and Fred McLeod Wilcox, "Forbidden Planet"*

In what sort of world, then, might we be living? How might we imagine it? In Figure 25, you can see a series of frames through which a timeline is randomly passing. I don't intend to imply that these frames are somehow suspended in a space and lined up in some fashion. Actually, what this figure is meant to imply is that there is one frame and that it is always ending and reforming its contents. The randomness of the time line is only meant to show that the contents of any frame are random and that these frames, therefore, appear in a random order.

Some of the frames are labeled with the letter A, which will stand for your consciousness. It is the patch of consciousness that has appeared within that frame at that instant. An unknown number is also associated with it, such as Ax to represent not only your consciousness (A), but also your memories, as they would exist at that point in your life (x). I argue that your memories are logical and make sense (e.g., you won't recall that your grandmother had four heads) because the operations of our universe, or frame, forbid memories inconsistent with the laws of physics that operate within that particular frame.*

---

*At this point you might wonder how one could explain a psychotic person who, in fact, is certain that his grandmother does have four heads. I would argue that it would be possible for you to perceive a four-headed grandmother so long as the laws of physics regarding psychosis in our universe were being followed. In a like manner, you might actually see that your grandmother has four heads (and believe it, too) if you were plugged into some sort of virtual reality machine. Again, such a machine would have to conform to the physical laws of our universe.

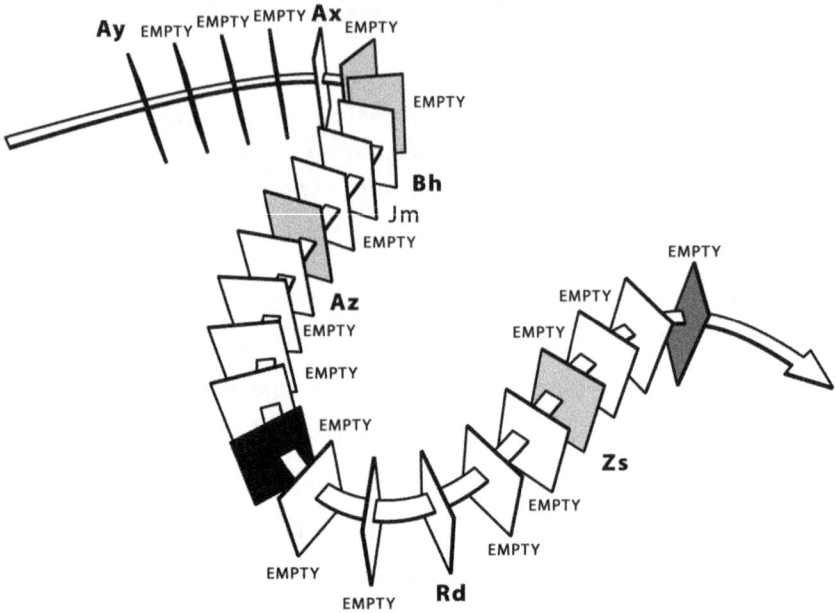

Figure 25. *Differing frames of space/time might appear, vanish, and reappear in a random fashion owing to the fact that absolute nothingness is, for some reason, unstable. Here we see an artificial timeline drawn through differing frames to represent the sequence in which the frames appear. A few contain a conscious mind at a particular point in that person's life, but most don't. Frames without a consciousness are labeled "Empty." This concept generates interesting questions. For instance, are persons A, B, R, and Z, really different people? Should frames that incorporate different laws of physics be rightly classified as different universes? (Frames containing the same laws of physics are shaded the same.)*

Most frames are labeled "Empty," as they contain no patch of consciousness, which is expected since the odds of things coming together to form a consciousness are presumed to be very remote. Frames that have other physical laws within them are in black and different shades of gray, as I am saying that they belong to another universe. Some of the frames are labeled B, C, or D along with various numbers. For the moment, let's say that these represent the memories of other people at various points in their lives.

If you look at the NOW represented in frame Ax, we might say that it is you at some point in your life. Perhaps we can say that it is right *now*. You exist, frozen in time, right at that point. Perhaps a single point in time, an instant, is unable to generate any self-awareness since time is required for the processes necessary to operate which can lead to a memory of thought. Perhaps frame Ax all by itself is not enough to enable "I think, therefore I am." Perhaps we will have to wait for the lucky random sequence Ax, Ay, Az to occur. In an infinite series, given enough time, such continuous strips of conscious-filled frames could be laid down. Perhaps at the moment such a strip appears (a series of closely related frames following one another), you become aware; you think, remember things, and are alive.

Frame Bh is interesting. Who is that? It would seem to be Consciousness B at point "h" in his or her life. But why is B different from A? If the only reason is that B has a different set of memories and perceptions associated with it, then A might not be different from B other than the "life" it is living. It might very well be, then, that B is also "you," only with different memories. Perhaps whenever the universe becomes conscious it is always YOU, and you are everyone. In other words, we are all the same person. In that case, what you are reading now is a message that you have written to yourself.

In such a universe there is no concern with the supervenience paradox, as there is only one consciousness. There is no concern with Zeno, as nothing is moving. There are no worries about Bell's Theorem because all is "here" and nothing is "there." Occam's Razor also applies as there is the minimal amount needed to yield the world that we experience. There is no Moon (unless you are looking at it). There is only your consciousness and perhaps a minimal perceptual and memory component...whatever can occur during the limited strip of consciousness that has lined up.

How would the world appear if all this were true? It would appear exactly as we see it every day of our lives. We would be blissfully unaware that anything was happening other than the usual. It might well be that a trillion "years" will pass between the time you read THESE FEW WORDS and the time you read THESE OTHER FEW WORDS. You would be none the wiser. In fact, you might well have read them in the opposite order and still you would never know.

So, in which world do we really live, the world with which we are all familiar, or the strange world of random, frozen, and endlessly cycling frames? Both worldviews have a certain amount of validity. Each one describes our existence as best we are able to understand it. They are, to a great extent, two sides of the same coin. In fact, if you like, you can argue that frames Aa through Az might actually occur in linear order (by random chance) allowing for the kind of life most of us imagine we lead, one that progresses in straight ordered fashion through a timeline from a "beginning" to an "end." Of course, as we have noted, whether the order were Aa, Ab, Ac, Ad, Ae, etc, or Ae, Ab, Aa, Ad, Ac, etc., it would be impossible for you to tell the difference. Either experience would appear to be the same. You may look at your world in either way. Both yield the same experience; only one view does so without also generating unpleasant paradoxes.

MULTIPLE WORLDS

For of all sad words of tongue or pen,
The saddest are these: "It might have been!"
—*John Greenleaf Whittier*

But now there are more difficulties, and they concern the imagined stability of frames. I have imagined that the frames that appear are stable in some meaningful way. I have proposed that they follow physical laws that are randomly unique to our universe. This prevents frames following upon frames from having wildly strange and disparate characteristics, such as the sudden appearance of a tentacle where your arm used to be, a violation of the lawful physics contained within the frame. I am arguing that physics, biology, chemistry, geology, etc., actually "exist" within these frames inasmuch as their lawful attributes will stay in force from frame to frame. All well and good, except for one small problem. Why wouldn't there be *lawful variations within* the boundaries set by these natural laws?

For instance, when I walk out to my car in the morning I expect to see certain things. There is my black car (it was the last model for that year so I bought it to save money, but trust me, never get a black car as it shows every speck of dust and dirt and it is really hot in the summer...but I digress). From my driveway I can also see the snow-capped Olympic Mountains off in the distance. If my existence is best described by a random series of frames capable of generating a consciousness and which follow strict natural laws, then why isn't my car sometimes red, or on occasion why aren't the mountains set a few miles further south? If my car were red, it wouldn't violate any laws of physics, or chemistry, or biology, etc. In fact, I believe that red cars are available for purchase, assuming that you fancy yourself to be some sort of fire chief.

Now I am in the position, if I wish to continue supporting the frames concept, to argue that whatever random (within lawful limits) force that yields up my consciousness also requires that the color of my car (whether it is "really" there or only in my mind) remain unchanging. Remember, each frame is independent of each other one. For them to all exist in what I am defining as "our universe" they must contain certain laws of physics that happen to continue by chance from one frame to the next (which maintains a series of frames within the confines of "our universe"—again, by chance). So why should the color of my car be preserved across frames that only require that the laws of physics be obeyed and not that *everything* in my frame remain utterly stable from frame to frame?

Oddly enough, quantum mechanics has an answer. It was first presented in 1957, in a Princeton doctoral dissertation by Hugh Everett, and has come to be called the "many worlds hypothesis." Everett was attempting to solve a problem that we have already discussed. It seems that atomic particles, such as electrons, exist in multiple states and that it is only our interacting with them that causes a particular particle to "collapse" abruptly and yield a single value. You might recall our photon analogy that had a coke can pop up out of the ocean near Lisbon. Had the *first interaction* been with a soda can in Iceland, *it* would have been the one upon which the entire wave collapsed to yield its force or value. In a similar fashion, electrons surround atoms, not as tiny moons in orbit about nuclei, but rather as "probability clouds" within which the electron is said to have a probability of residing at any given point. In fact, there is a probability, albeit a very rare one that an electron associated with the nail at the end of your little finger is out near Mars at the moment. To be honest about it, there is a probability that every particle in your entire body, and in the arrangement that it currently maintains with every other particle in your body, might suddenly reconvene itself on Mars. The laws of quantum physics actually yield a real chance that you might just suddenly "leave" where you are and "appear" on Mars (although it would be more likely that just a hand or foot would make the trip). Physicist Michio Kaku says that he likes to have graduate students actually calculate the chances of such a sudden trip as part of their homework. (I can't help but wonder if the students who make errors and come up with a high probability don't lose sleep worrying about suddenly having a Martian address). Fortunately, the odds against this are so wildly high that you needn't concern yourself with that unsightly prospect. (Although the odds are a good bit better that you would just scatter randomly about your house or apartment—but the odds are pretty unlikely on that account as well). This bizarre possibility comes about because all things seem to be "everywhere" until an interaction resolves where they "actually are." In other words, the probability cloud of an electron does not resolve itself as an actual electron at a given point until we interact with it in some way. We have also seen through Bell's Theorem how our interactions appear to force consequences at a "spooky" distance simply by our interacting with the process.

This whole thing was becoming something of a quandary and Hugh Everett was set on overcoming it. Everett saw that physicists were struggling with the clear implication of quantum mechanics that shows there to be multiple states for every existing particle. To explain how there is only one "actual" state of existence, namely, the world that we see around us, physicists relied on the "collapse model" in which, as we have stated, multiple states collapse into one state as soon as we interact with them. This, of course, argues that physics as we know it can't exist until consciousness becomes part of the equation! Having to add consciousness as a variable to every physics equation was causing many physicists to throw up their hands and cry out, in the words of Oliver Hardy, "Here's another fine mess you've gotten me into."

Everett, however, had a breathtaking way out, and one that squared with the findings yielded from quantum mechanics. He argued that *multiple states were real!* All possible variations of a probability cloud were really "there," and if we only saw one of the many states it was because other versions of ourselves were experiencing the other probabilities in different parallel universes! According to this view, whenever a probability is resolved, it creates a distinct and separate universe. So, in one universe my car is black, in another it is red, in another it is black and the Olympic Mountains are a few inches further south, in another it is blue and the mountains are a few more inches further north and my mail box is a centimeter wider, etc.

Everett's dissertation fell on deaf years and was basically ignored. Even though it explained one of the greatest mysteries in physics, it was over-looked, probably because the idea of *real* multiple worlds was too breathtaking to be embraced. Until his death in 1982, Everett continued to argue that people shouldn't be afraid to take a good look at what the mathematics were telling them, and that multiple worlds were real.

Physicist Bryce DeWitt, one of the few exceptions who saw Everett's work early on as seminal, took an interest. DeWitt tried to understand physicists' reluctance to accept the multiple worlds hypothesis. DeWitt said, "Of course, the number of parallel universes is really huge. I like to say that some physicists are comfortable with little huge numbers, but not with big huge numbers."

Perhaps DeWitt is correct. The idea that there is, for instance, a real electron at each location within its probability cloud, just in a different universe for each location, offers an awe-inspiring conceptualization. One must be willing to believe that if, for instance, a physicist measures a photon so that he or she might decide if it is acting like a particle or a wave, the universe would divide, making exact copies of itself, including copies of the laboratory and the experimenter. In one universe a particle would be observed, in the other, a wave. A photon might act at any given instant like a particle or a wave, just never both in the same universe and at the same time.

When thinking about the many worlds hypothesis, the number of possible worlds is truly overwhelming. If you think about how many parallel universes would have to be created by the action of a single photon climbing out of a star, it leads to a view that is, in the words of theorist Philip Pearle, "uneconomical." And yet, other physicists, such as David Deutsch, a fellow of the British Computer Society and perhaps the current leading advocate of the many worlds view, argues that infinity is after all...huge. As Bryce DeWitt noted, huge numbers are intimidating, and infinity is as huge as huge gets.

Some physicists would rather not dwell on the prospect of the many worlds hypothesis. Steven Weinberg, one of the most eminent of living physicists, compares the hypothesis with Churchill's view of democracy. Churchill said that democracy was the worst of political systems, save for all the others. When pressed about what he thinks of the multiple worlds hypothesis, Weinberg has been quoted as saying, "I don't know...I think I come out with the pragmatic people who say, 'Oh, to hell with it. I'm too busy.'" However mind jolting it might be, the problems and paradoxes associated with our observations at the quantum level disappear if the multiple worlds view is accepted. It would also nicely explain the variability that one would expect to occur within frames.

Once again, imagine a frame in which there is conscious awareness that has been created (you). This frame is in *our* universe and therefore is constrained by the physical laws that have come to define our universe. In other words, things can change from frame to frame, but only things that conform to the physical laws of the universe will be available for the consciousness to perceive.

Now, imagine a stable frame strip that has yielded your consciousness. Among the few perceptions you have in this strip is the appearance of a green car that you recognize, by checking your memory (which is a linked to the perception you have of your car, perhaps even generated by the perception of your car, so that it recalls that your car is green).

If in the next strip of frames that happen to come together your car is yellow, any memories that you might access of your car would be that you have a yellow car. In this instance, *you are in another of the multiple worlds* that Everett proposed. You can make up any number of combinations you like, car color, weather that day, what your neighbor says as you pass by, etc., and each should be internally consistent with the allowable memories you access and the world you perceive. The idea of frame strips, I believe, makes the multiple worlds view a bit more palatable. I also would argue that whenever another world is created, it should not rightly be considered to be a different universe. I would reserve "different universe" to describe a place in which frames are created according to different universal laws. I would assume that such universes would most often be places in which no consciousness would ever be generated simply because no physical law in that universe allowed for it.

In this view, the multiple worlds are all within one universe. The differing worlds are comprised of differing frame strips that generate a consciousness. Worlds differ when frame strips differ through random variations (within the boundaries of the physical law of the universe in which frames are being produced). In other words, in one world your car is yellow and stays yellow unless you paint it, while in another it might be green (and, similarly would stay green unless you painted it). Multiple worlds, therefore, would be the result of all possible lawful variations (in some worlds World War II never occurred, or the tallest mountain is something other than Everest, or any other possible combination of just about anything you could imagine), and each individual world would be the result of consistency between your memory and your perception according to the lawful expression of the physics of our universe (green cars stay green unless painted, and Everest stays the tallest mountain unless there is a geological reason for it to change). Of course, the physics of cars and the laws of geology need only be represented within your mind as a manifestation of the physical laws governing the creation of your consciousness. If it's all "in your head." then there can actually be geological laws without the need for any actual geological structures (even Everest isn't there unless you are in the process of perceiving it). In other words, if your consciousness in a particular frame strip perceives a geological law, then the *law* might exist even though there might not be any actual geology to go along with it.

It is also interesting to consider that in these many worlds (i.e., stable frame strip variations) you might look a little different. In one you might be an inch taller, in another you might have a slightly different nose. Your memories might also be slightly different in each of these worlds. In one, you recall that your car is yellow, in another, green. There are lots of different variations of you throughout the many worlds. Perhaps in some, you look quite a bit different, perhaps enough so that if you could see a photo of yourself from that world, you wouldn't recognize yourself. Perhaps your memories would be fairly different, too. I ask you, then, at what point are you no longer you? At what point are you somebody else?

As you can see, the idea of many worlds also blurs the line that separates *you* from *others*. This is why I would argue that whenever consciousness is generated, it is you. We are all the same person. I can't prove it, not in any empirical sense, but it logically fits with what we know. In multiple worlds, in which there resides only one consciousness, there is simply no clear place to draw a line that separates you from others.

It is interesting that at this point we have also left our "bodies" behind. Instead of conceiving of consciousness as existing inside of the brain, we are considering that the brain exists inside of consciousness (in the same way that the moon might, or might not, depending on whether you were recalling or perceiving it). There is no longer any value in trying to hunt for "you" inside of the convolutions of your brain, because your brain is a part of your consciousness, not the other way around. In a frame strip in which a brain "existed," there would be a consciousness and memories or perceptions of a "brain." Whether this consciousness, like Aristotle, perceives the brain to be a "radiator," or whether it has a more sophisticated understanding, is of no consequence. The laws of physics governing the universe in which the frame strip was created would only limit one to memories or perceptions that were *possible,* not necessarily ones which were "correct."

« »

At this point I wish to depart from the text for just a bit to conduct a brief review of the main concept I am developing by giving an example.

For a moment I am going to close my eyes, place my elbows on my desk, and rest my head in my hands. As I do this I happen to notice that the back of my left hand itches a bit.

Now I want to explain the moment I just described in terms of our discussion.

Absolute nothingness is unstable and a frame (a particular cycle of "sparks") appears. This "frame" is the entire universe...it is everything. In an infinite series (and I am assuming these frames cycle into existence forever) all non-zero events must occur, and occur infinitely. The frame that has suddenly appeared has the physical laws (generated at random) that we associate with the laws of "our universe." Also, in this instance, the "sparks" so arrange themselves that this universe will become aware of itself and become conscious if nearly identical frames appear in a sequence (here we introduce time as a crucial element—but perhaps it is not needed). The series of frames need not be exactly alike, but close enough (within the parameters set by the uncertainty principle, keeping the series of frames from becoming too "slippery" to generate a consciousness). The fact that consciousness might be generated by frames that are perfectly matching, but are close enough in sequence, might well explain the underlying uncertainty we see in our universe. When such a fantastically unlikely series occurs (like random snow on a television set suddenly creating a picture) the universe becomes self-aware. As a non-zero event, such a sequence must occur over infinite cycling no matter how unlikely it may seem.

The universe, in this case "I," feels certain pressures, but sees nothing (I am resting my head in my hands and have my eyes closed). However, unless I am thinking of my eyes or the rest of my body, there is no need to consider that these aspects exist in this frame (universe). It is a universe of some felt pressure and awareness (the chair against my body, my hand against my head, etc., an itch, a thought about the itch, a sense of breathing), but nothing else. There is no way to know on which planet this is occurring, what the "chair" is, and no way to know what sort of life I am leading, whether I am a man, a woman, something else...not in this frame series. The consciousness suddenly ends when the frame cycles into something with perhaps different physical laws and no consciousness. The consciousness might not appear again for a trillion "years" of random frame cycling.

What was just experienced by me (the universe) was a brief moment in one of the lives in one of the many worlds. Who knows what it was, what it meant...it doesn't matter. It was a blip. Like the sudden appearance of a few seconds of a real TV show popping out of the random "snow" for just an instant. This is what I am suggesting life is...all our lives. It is little blips here and there, all possible blips, most very short, a few quite long in a random and endless cycle of nothingness bumping about because it is unstable. We can remember the arc of our one life in the same way that we can remember an entire TV program even though there was never more than a dot on the screen at any given time. Full lives lived in sequence are an illusion, albeit a very powerful one. Considering life to be endless blips of all varieties will yield the same experience you are having right now, but without all the paradoxes, and with the whole business being far simpler than we generally imagine. The commonplace view of a life with which we are all familiar requires far more complex and inexplicable actions and laws than this new and simpler view, a view tightly in line with modern physics, astronomy, and psychology. This new understanding also indicates that you are immortal, alone in the universe, and will eventually have every experience and be every sentient being that has a non-zero probability of occurring.

WHERE DO WE GO FROM HERE?

There was never a time when I did not exist,
nor you, nor any king of the world.
Nor will there be any future in which we shall cease to be.
                                        —*The Bhagavad Gita*

It is time now to return to the great question, "What happens to us after we die?" Clearly, we cannot know. But I will venture a guess based on what it appears that we do know. I am guessing that when we die, we simply become someone else. By "someone else" I mean a sentient being capable of conscious thought residing on a planet capable of sustaining such life. There is no need to become that person at the start of that person's life when he or she was born, you might happily drop in anywhere in that particular life span.

But if it is someone else, is it you? I believe it is; as we discussed before, it would be you, but with memories of an entirely different life. But when that person saw something funny, it would be you who laughed, if that person was stuck with a pin, it would be you who felt the pain, and if that person was loved, it would be you who felt the comfort and joy. So I will argue since you seem now to be *living that person's life*, it is you.

Now if you ask me, do I really believe all of this, I might hesitate. If I want to be painfully honest with you, I would guess that what happens after we die might well be something that no one has ever considered. There might well be possibilities unknown to us. In fact, I am sure that there are. The answer might lie out there, in the unknown. But for now, my best guess based on reason, logic, and empirical evidence is that we just become someone else.

The primary purpose of this book, however, was not to just give you my best guess, but rather to attack the commonly held secular assumption that when we're dead, we're dead, and that's that. I can't say for sure that

that particular view is incorrect, but I think by now you can see that it is beginning to look like a very improbable outcome. You needn't accept the idea of frozen universes or random frames, either. You can hold back within the boundaries of what empirical science can prove and come to the same conclusion.

Let's briefly go back over what we have discovered and examine the assumption that "when we're dead, we're dead, and that's that." You are alive and conscious (I think, therefore I am). Something, probably some combination of things, has led to the creation of you in this universe (you actually do exist). Empirically, we know that there are more stars in the known universe than there are grains of sand on all the beaches in the world and that most of these stars probably have planets in orbit about them. We also know that the stars and planets in the universe appear to be made of the same elements that we find in our own star and planet. Is it reasonable to assume that there is some strange rule that says whatever led to the creation of your consciousness can never happen again in that stunningly vast universe, ever?

The great majority of astronomers and cosmologists have now concluded that our universe is but one of many, perhaps an infinite number of universes. There is also an indication that our own universe might be hugely larger than what we can currently observe. Such prospects only make the assumption of a repeat creation of you seem all the more likely.

If you go further and accept the other things we discussed, I believe the prospects grow stronger still that you do not simply die and that's that. But perhaps you find the ideas of frames and frozen universes and nothing existing unless you're looking at it a bit too strange to believe. I agree with you. I live my life day to day, as do you. The people and things I see, and with which I interact, appear very, very real to me. The moon certainly looks very real, and I can't believe that it is just a figment of my imagination. That's the thing about the present, or "the tyranny of the present," as it is sometimes called. It way outclasses recollections of the past or dreams of the future for "realness." I can imagine a car accident in the future, or even recall one from the past, but neither is anything like actually being in the middle of one while it is occurring. Reality is, well.....real.

However, I can't help but to recall an orrery I once saw. If you are unfamiliar with the word, an orrery is a planetary model driven by gears or clockwork so as to simulate the motion of the planets in the sky. When I was a boy, I recall that the Hayden Planetarium in New York City had an orrery on the ceiling of the room in which you waited before entering the adjoining room for the star show. On the ceiling was a yellow globe to represent the sun. Circling about the globe on tracks were the planets, many with individual moons in orbit about them. It was all done with gears and the planets and moons moved in a roughly proper relationship to one another. Mercury would zip about the sun, while further out Saturn would take its time.

But the orrery that now comes to mind, the one I want to tell you about, was not the one at the Hayden Planetarium. The one I saw was very old and showed the Earth at the center of the solar system. This particular orrery was fantastically complex because it had to make many special things happen in order for the motion of the planets to be true to observations while maintaining the Earth at the center. For example, Mars would move along in its orbit about the Earth, and then suddenly stop, back up for a distance, stop once again, and finally proceed forward (see Figure 26).

Looking at Figure 27, you see that the movement of the Earth about the sun causes this apparent retrograde motion. In this example Jupiter isn't actually reversing motion, but only appears to do so when observed from a much faster moving Earth as the Earth overtakes it and passes it by.

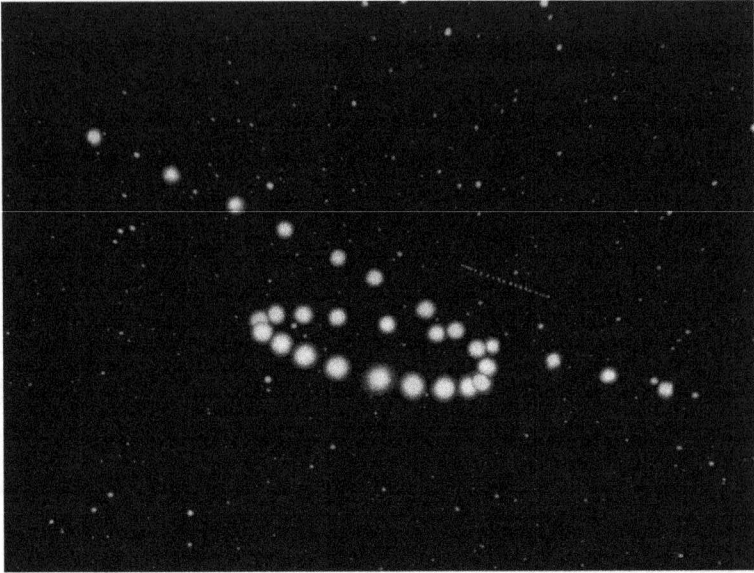

Figure 26. In this image, you see a series of photos of the planet Mars taken on successive nights from the same location on Earth. Over time, Mars appears to move backwards across the sky before, once again, resuming a forward motion. Such apparent retrograde motion can be observed in Jupiter and the other outer planets as well, and is due to the Earth's greater orbital speed.

Credit & Copyright: Tunc Tezel

When the sun was finally placed in the center of our solar system things got a lot easier for orrery makers. Many less gears were required and the process was made far less complex (Occam's Razor, once again). Not only were orreries now simpler to build, but also the paradox of retrograde motion was gone.

So I look about me at the very real world, and then read about the sorts of paradoxes we have discussed. I think of Occam's Razor and of the fantastically complex orrery.

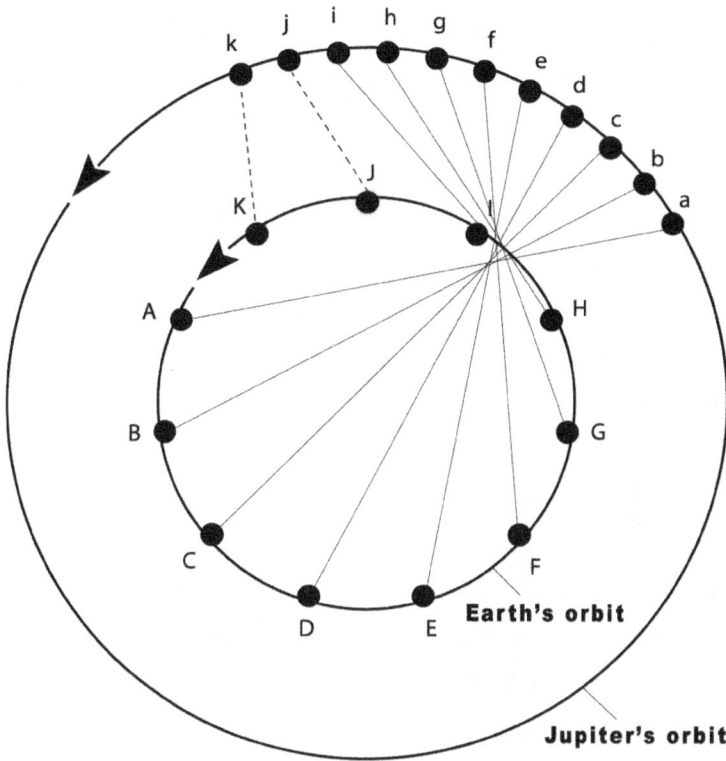

Figure 27. The Earth will make more than 11 orbits of the sun in the time it will take Jupiter to make one. In the above example, Jupiter is sighted from Earth throughout an entire year. When observed over time from positions A-J, Jupiter will appear to progress "forward" in the night sky. But when the Earth moves from viewing position J to K (dashed lines), Jupiter will appear to briefly move backwards. This apparent retrograde motion is the result of Earth "overtaking" Jupiter in the race around the sun. Once the laws of gravity were understood, retrograde motion could only be explained by placing the sun at the center of the solar system. (In this figure, the radius of the Earth's orbit is greatly enlarged to make the diagram easier to inspect. In reality, the degrees of viewing angle would be far less.)

I also think of how very often rock-solid common sense has turned out to be dead wrong, and I am drawn to the idea of a simpler world in which the paradoxes we have discussed are resolved; a world that would be indistinguishable from the world in which you and I live. I think that quantum physics seems weird to us in the same way that placing the sun at the center of the solar system once seemed weird. I think that this time, too, we should listen to what the numbers are telling us.

Why make things more complex than they are, just because they look that complex owing to our limited understanding? I think that there probably are multiple universes, and that in ours we live in a series of random frozen frames. Maybe we do, maybe we don't. But it seems reasonable to me. Either way, the many worlds hypothesis would indicate that you have an infinite number of lives to live. Every possible variation of your life is there. Whether you are also the "other people" in the world, I don't know, but it seems like a good possibility. Or perhaps there are different domains, as philosophers like to call them, limiting you to being you and a certain collection of "other people" but not all of them. Then again, if we suppose that, we will once again get into the "what makes them different from you" problem, and I would rather employ Occam's Razor and go with the simpler solution. The point is, "when you're dead, you're dead, and that's that" now becomes a possibility that appears extraordinarily odd and not very likely. To postulate such a thing would probably require rules and speculations far more outlandish than any I have offered for this new view. Oh, I suppose you might say that absolute nothingness became unstable just this once, and now that it has gotten this one belch out it will happily revert to not being there or anywhere and take "there" and "anywhere" with it when it goes "nowhere" and is therefore gone from no place that would then never again exist (and actually never existed since time would have left and gone nowhere at the same time that nothing did). You must admit that such behavior does seem a bit capricious. I prefer to believe that you can count on "nothing" (dreadful pun, I know, but irresistible).

DO UNTO OTHERS...

He hoped and prayed that there wasn't an afterlife.
Then he realized there was a contradiction involved here
and merely hoped that there wasn't an afterlife.
—*Douglas Adams*

Suppose that it is true that we never actually die, in the sense that we never come to a permanent end? Suppose we do lead, not just our own lives, but all the lives of everyone else as well, and for added good measure, all the possible lives that the many worlds hypothesis offers?

At first glance it might be a relief to consider that death is an impossibility. It is perhaps comforting to think that we simply get to continue, albeit along a variant of either ourselves or ourselves as "someone else." Furthermore, if we happen to be in love with our current life, it is good to realize that it, too, as one of the variants, might be relived endless times, since there is nothing that we are imagining that would forbid repeats of any particular frame strip.

However, the whole idea of endless lives becomes less comforting when you stop to consider the history of human beings. How many have led happy lives as opposed to those who have led miserable, brutish, lives; lives filled with agony and pain? I don't know about you, but I don't want to be a slave, or someone who ends up starving to death after watching his family die, or a 19-year-old who gets cut to bits by machinegun fire in the First World War or some similar war on some other planet. I can think of lots worse things, too, and I am sure that you can as well. The problem seems to be that what's really bad is a lot worse than what's really good.

Think about it. Imagine a "happiness meter" (sort of like an applause meter) with the needle resting midway at zero. If the needle moves into positive territory, to the right, it means that your experience is a good one:

The farther to the right, the better the experience. Conversely, if the needle moved to the left it would mean a bad experience, the farther to the left, the worse the experience. If we had to place numbers on this scale, we might say that the good scale goes all the way to +10. +10 is the very best possible experience. However, if you were honest about it, wouldn't you agree that the bad scale ought to go all the way down to about –100?

I don't know what a +10 would be for you, perhaps winning a billion dollar lotto, or something else that you have fantasized about in your wildest dreams. But I would be willing to argue that being tortured to death (like at the end of the film "Braveheart," in which the hero is strapped down in front of a cheering mob while some hairy barrel-chested guy wearing an executioner's mask slices and dices him up with gleaming, peculiarly shaped implements, that enable him to inflict the most agony) wouldn't be thought of as the exact opposite of winning a big lotto. I would not consider that to be a sort of yin and yang thing. One might be a plus ten, but the other has to be down there in the minus nineties somewhere.

Of course that is in this world and at this time. Perhaps in a world where heaven loops are available every instant would be a +100 with nothing ever in the minus column. And there might be better worlds than this with happiness and comfort at far higher degrees than anguish and pain.

I admit part of my exploration of this topic was an effort to find a way out of having to die. As my mother used to tell me, "Just because all the others are doing it, that doesn't mean that you have to." But I fear that I might have discovered something worse than death...life. As I said earlier in this book, I didn't mind being dead when I was "dead," so perhaps one life, lived only once, is preferred. In the words of Miguel de Unamuno, "And even if we were to succeed in imagining personal immortality, might we not feel it to be something no less terrible than its negation?" Maybe we should hope that we die and then get left alone. If it is the other way, and it looks as though it is, then not only will bad things happen to us, but every horrible thing that has ever happened to anyone will happen to us, and (when considering multiple worlds) every terrible thing that ever *could* happen to us will happen to us. Of course, that would go for the good as well as the bad. As Neill Hicks said, "Space is what keeps everything from happening to you." If we consider Zeno, and agree that there is no *there* there, but only a *here* here, in other words, there is no "space" beyond one tiny point, then Neill Hicks is right, and everything *will* happen to you! This conceptualization fits well with the many worlds hypothesis, because that is exactly what the hypothesis implies, that *everything* will eventually happen to you. However, I worry that the bad things often plumb depths that the good things have little hope of counter-balancing. How then, can one hope to deal with true horror so that the prospect of endless lives might be considered to be a good thing?

One way, might be to try as best we can to look on the bright side, although trying to find the bright side of misery, suffering, torture, and multiple horrible deaths might be a bit of a challenge. Let's start with the fact that a very horrible life, filled with misery and suffering is most likely to be a short life. At least we can hope so. There is also the added benefit of entirely losing one's memory when moving on to another life. So perhaps a good bit of the trauma of misery, anguish, suffering, and agony, are the *recollections* of it all. You and I are alive now, and if what we are supposing is correct, then we have lived a multitude of times before, presumably with numerous occasions of horror. And yet I am not less willing to live my current life because of that, and I dare say that you aren't either. Perhaps you or I, then, could get through being tortured to death just so long as we knew nothing about it afterwards. This, by the way, is what I am referring to as "looking on the bright side."

Another thought that might give some comfort is the realization that if you lose someone you love very much, you will have many other lives with that person, lives in which that person is not lost to you. If what we have been discussing in this book is true, then the death or loss of another is never a permanent condition. If we are all the same person, and if we can never die, then "others" cannot die as well. In fact, when you see other people, it might not be incorrect to think of them as sneak previews of a life you will someday lead (along with every possible variation, even those you don't get to see in this life).

Oddly, population demographics can also provide some comfort. When I was a boy, the population of the United States was 140 million. Now it is about 300 million. A far more amazing statistic is that one-quarter of all the people who have ever lived are alive today. This is because advances in technology have made it possible to support more people than ever before. If humans still lived a nomadic life, the entire planet could probably support little more than the current population of California. But modern agriculture, medicine, and social advances enable the support of a far greater population.

These data tell us that worlds with very large populations are most likely to be good ones in terms of health, safety, comfort and protection. Worlds without war and well advanced scientifically are more likely to have extremely large populations and long-lived ones at that. Worlds in which you would not wish to reside are the most likely to grant you your wish, and probably sooner rather than later. In terms of probability, then, the chances that you will experience a number of very good and fulfilling lives without horror and hardship may be a good bit greater than we might have first assumed. I, for one, am looking forward to being a demigod and playing out every happy dream that I might imagine while safely cocooned away below my home world's surface. Perhaps it is worth living through the bad stuff to get to that.

Curiously, there is also a moral dimension to the world I have introduced to you that is a bit troubling. If you are the only consciousness in the universe, and all the "people" around you are not "real" in the sense that no other consciousnesses reside within them, then it stands to reason that you could do whatever you wished to those people without it having any real consequences. You might say, "Well, if I kill someone I will likely be caught and sent to prison." But in the world we have come to understand, a trip to prison is simply an event that you will most likely experience in one of the multiple worlds regardless of what you have done, and doing away with someone would have nothing to do with it, although your memory and sense of continuity would make it seem otherwise.

This observation has been troubling to philosophers and physicists for many years, even before modern physics led to the conclusion that there must be multiple worlds and universes. It was debated back when such a possibility was thought to be even remotely true, and was found to be deeply disturbing. The first to discuss the problem of morality as it would pertain to in an infinite series of worlds and selves was Friedrich Nietzsche in *The Will to Strength* written in 1886. Nietzsche wrote:

> The universe must go through a calculable number of combinations in the great game of chance which constitutes its existence... In infinity, at some moment or other, every possible combination must once have been realized; not only this, but it must also have been realized an infinite number of times.

Nietzsche drew this conclusion from the consideration that time, rather than space, might be infinite, but it yields the same argument. It troubles me that if it is "all in your head" and all experiences will eventually come your way that nothing you do to others actually matters. What value would there be in morality? If both good and evil are infinite in an infinite universe than nothing you do can hope to add or detract good or evil from it.

I once traveled to Israel and spent some time there with a friend. He was a fine man and a well-respected physician. At the time I was running low on ready cash and remarked in his presence that I was spending more than I had intended and that if I kept it up I would have to rob a bank. I know that he knew I was joking, but his response was interesting. He said, "You don't want to rob a bank." I asked, "Why not?" He responded, "Because you wouldn't enjoy it."

I thought about it, and he was right. Even if I were wildly desperate for money I would hate the entire experience of actually robbing a bank so much that I could never go through with it...never mind that it was the wrong thing to do or that I might get hurt trying it. The answer that he had given me was actually quite profound. At one level you might say that to not rob a bank because it wouldn't give you pleasure belies a fairly low level of moral development, an almost hedonistic method of determining right and wrong. But if the reason that I would find robbing a bank so repugnant was because I had a strongly developed sense of right and wrong, especially in terms of how my actions might affect others, then not acting because it would displease me would, in fact, indicate a fairly developed sense of morality.

I would argue, then, that I live in a moral universe (at least at this moment in my life) because I would find such an action repulsive. Even if I don't believe that "other people" are real, I don't care to see a preview of what will eventually be happening to me when I become those others by putting on a show in front of my own eyes by doing something evil. If such thoughts and feelings are in this frame strip, then my universe is currently a moral one. Of course, there are probably many frame strips that are amoral, or immoral. Morality might well be a random thing. At least at the moment, being immoral would make me unhappy. Or as Einstein said while pondering this problem, "I know philosophically a murderer is not responsible for his crimes, but I prefer not to take tea with him."

I find such thoughts about morality disturbing, to say the least. But I have no intention of going through the rest of my life treating people as though they are not "real," even if I rationally suspect that they are not. Frankly, I find the whole thing too spooky. Then again, spookiness appears to be a mainstay of quantum mechanics.

If other peoples' lives are previews of ones I will be living, then it might be best that I treat other people as well as I possibly can. I say this for selfish reasons, but it does appeal to my sense of justice. I say this because it would mean that whatever you do to "other" people will sooner or later be done to you (by you) since there will come a time when you will be that person. If there was ever a reason to do onto others, that would be it. Considering our bank robbing example we must ask if the net evil done to all affected by the robbery will outweigh the good? If you rob a bank, and you would eventually be everyone else affected by your actions, I would argue that YOU will suffer a net loss by your bank robbing decision.

If such a world were the way of things, there would also be other curious moral twists. If I saw someone who was homeless and suffering, I might rationally think that I will hate that life when I get to it, and that the best thing for me to do would be to end that life quickly and send "myself" on to something better. Is it murder, or giving yourself a helping hand? Could, "If you ever find me in a situation like that, just shoot me," be a message to yourself? And would it do any good to shoot "yourself" since on other multiple worlds you will decide not to shoot yourself? If, owing to the multiple worlds hypothesis good actions and evil actions are infinite, than no action you take can add or detract from good or evil and morality becomes a moot issue.

Many of the moral implications of our discussion are upsetting, if not appalling. But even that should not deter us from following the logic. No one says that the universe needs to follow a certain socially proper path. But I am not advocating murdering anyone who seems unhappy (an *Arsenic and Old Lace* philosophy) or committing suicide if things aren't all that good in this life (something akin to Philip José Farmer's *Riverworld*). I say this because of something Jacob Bronowski once said.

In his book *The Ascent of Man*, Bronowski discussed a visit he made to Poland to see the remains of the Nazi death camp at Auschwitz. He wandered about the grounds near the crematorium and scooped up some dirt with his hands. In that dirt he knew were mixed in the ashes of the dead. (So many had died at Auschwitz that no patch of ground was devoid of such remains.) He held the earth in his hands and asked what sort of people could do such a thing to other human beings? He concluded that the Nazis were capable of such action because their beliefs left no room for doubt. He concluded that if people are absolutely certain that they are right, they become capable of any action.

EPILOGUE

When I completed the first draft of this book I was feeling perfectly well. In fact, I still feel perfectly well. But by the time you read this book I will most likely have died.

I had no symptoms whatsoever of the kidney cancer that is about to take my life. It was discovered quite by accident while checking me out for something else that I didn't have. And let me tell you, this is rough stuff. I don't want to die and I have teenage sons living at home and a wonderful wife. I want to stay with them and make sure that they are okay.

So now I look at this book I have written in a new light, and find strangely that I am taking comfort from it. Even though I have described a universe in which we are immortal, alone, amoral, and subject to pain, anguish, and pleasure that is not under our control, I take comfort in knowing that everything is probably not coming to an end; that there will be times when I will go on to be with my family. In fact, I will in turn be each of them (which makes me glad that my wife and I bought the boys their own computers when they asked for them). There won't be a last sunset; a last kiss good-bye.

Of course, there might be lives filled with pain and suffering to pass through before I find happiness again, but I don't care. Bad memories get erased as well as good ones.

I admit to being very frightened, but since this book has given me considerable comfort, I find myself hoping that it might do the same for others. I now see death, not as some black pit into which I am about to fall, but rather as a great adventure that will eventually lead me back to all those I have loved and lost in every life.

Like I said, by the time you read this I will most likely be gone. And yet, I think I know what I am doing right at this very moment. I am you, and I am reading this sentence.

INDEX

ABOUT THE AUTHOR

John Dworetzky was born in Cooperstown, New York, on August 14, 1947 and died in Bellingham Washington on February 26, 2008, shortly after completing this book. While he was still a child, his mother predicted that he would grow up to become either a teacher or a writer; John fulfilled her prediction by becoming both. He obtained his Ph.D. in experimental child psychology from Utah State University in 1975, and went on to author three college textbooks that were published in many editions for West and Wadsworth Publishing Companies. He taught introductory psychology and child development courses at Glendale Community College in Arizona and when he moved with his family to Bellingham, Washington, taught in the Psychology Department at Western Washington University.

www.ingramcontent.com/pod-product-compliance
Lightning Source LLC
Chambersburg PA
CBHW031244090426
42742CB00007B/303